SDG 2 – ZERO HUNGER

CONCISE GUIDES TO THE UNITED NATIONS SUSTAINABLE DEVELOPMENT GOALS

Series Editors
Walter Leal Filho
World Sustainable Development Research and Transfer Centre, Hamburg University of Applied Sciences

Mark Mifsud
Centre for Environmental Education and Research, University of Malta

This series comprises 17 short books, each examining one of the UN Sustainable Development Goals (SDGs).

The series provides an integrated assessment of the SDGs from an economic, social, environmental and cultural perspective. Books in the series critically analyse and assess the SDGs from a multi-disciplinary and a multi-regional standpoint, with each title demonstrating innovation in theoretical and empirical analysis, methodology and application of the SDG concerned.

Titles in this series have a particular focus on the means to implement the SDGs, and each one includes a short introduction to the SDG in question along with a synopsis of their implications on the economic, social, environmental and cultural domains.

SDG 2 – ZERO HUNGER: FOOD SECURITY, IMPROVED NUTRITION AND SUSTAINABLE AGRICULTURE

AMBE EMMANUEL CHEO
United Nations University, Germany

KUGEDERA ANDREW TAPIWA
Great Zimbabwe University, Zimbabwe

emerald PUBLISHING

United Kingdom – North America – Japan – India
Malaysia – China

Emerald Publishing Limited
Howard House, Wagon Lane, Bingley BD16 1WA, UK

First edition 2021

British Library Cataloguing in Publication Data
A catalogue record for this book is available from the British Library

ISBN: 978-1-78973-806-3 (Print)
ISBN: 978-1-78973-803-2 (Online)
ISBN: 978-1-78973-805-6 (Epub)

Printed and bound by CPI Group (UK) Ltd, Croydon, CR0 4YY

ISOQAR certified Management System, awarded to Emerald for adherence to Environmental standard ISO 14001:2004.

ISOQAR REGISTERED
Certificate Number 1985
ISO 14001

INVESTOR IN PEOPLE

CONTENTS

LIST OF TABLES

CHAPTER 3

CHAPTER 4

ABOUT THE AUTHORS

Dr Ambe Emmanuel Cheo holds a PhD in Environmental and Resource Management (ERM) from the Brandenburg University of Technology Cottbus-Senftenberg, Germany. His PhD thesis *Integrated Water Resources Management (IWRM): Case study of the Far-North region, Cameroon* was published as monographs by the Verlag Dr Kovač. In his earlier research studies, he focussed on popularising the implications of climate change on water resources to policymakers, stakeholders, and vulnerable farmers, with the aim of influencing the decisions they make. Currently, he is working with the United Nations University – Institute for Environment and Human Security where he is contributing to research activities, capacity building and outreach activities within the framework of projects at the Institute.

Kugedera Andrew Tapiwa holds a Postgraduate Diploma in Education from Zimbabwe Open University, an MSc in Agroforestry from Bindura University of Science Education, a BSc Honours Degree in Agriculture from the University of Zimbabwe, and a Certificate in Pig Production from the Pig Industry Board. He is a DPhil student in Land Management and Conservation and lectures on Agriculture courses at Zimbabwe Open University and Great Zimbabwe University, located in Masvingo town, where he lectures on Agriculture courses, supervises undergraduate research projects, and assists the universities in fulfilling relationships with local communities and supporting them with agricultural knowledge to increase food security. He has vast experience in agriculture, agroforestry and land management and is presently working on rainwater harvesting and agroforestry research in arid and semi-arid areas of Zimbabwe to improve sorghum productivity.

LIST OF ABBREVIATIONS

Agriculture Orientation Index	AOI
Community Supported Agriculture	CSA
European Union	EU
Food and Agriculture Organisation	FAO
Food Insecurity Experience Scale	FIES
Gross Domestic Product	GDP
High-level Political Forum on Sustainable Development	HLPF
Inter-agency and Expert Group on SDG Indicators	IAEG-SDGs
International Centre for Trade and Sustainable Development	ICTSD
International Labour Organisation	ILO
Non-communicable Diseases	NCDs
North American Free Trade Agreement	NAFTA
Public Distribution System	PDS
NCD Risk Factor Collaboration	NCD-RisC
Southern African Development Committee	SADC
Sustainable Development Goals	SDGs
Sustainable Development Goals Fund	SDG Fund
United Nation – Division for Sustainable Development	UN-DESA
United Nation Economic and Social Council	UNESCO
United Nations	UN

United Nations Department of Economic and Social Affairs	UN DESA
United Nations Statistics Division	UNSD
World Food Programme	WFP
World Health Organisation	WHO
World Trade Organisation	WTO

1

INTRODUCTION TO SDG-2

The sustainable development goals (SDGs) address global challenges that are crucial for human and environmental survival and set critical thresholds for the use of natural resources now and in the future. The key feature of the 2030 Agenda for Sustainable Development lies in its universality and indivisibility (Barbier & Burgess, 2017; World Bank, 2017). The goals recognise interlinkages and foster strategies that build economic development and address social needs such as education, health, social protection, and job opportunities while tackling climate change and environmental protection. The achievement of the SDGs will require the participation of everyone (governments, the private sector, civil society, and every human being across the world) and a profound transformation of how humans think and act.

SDG-2 is one of the 17 SDGs and aims to end hunger, improve nutrition, promote sustainable agriculture, and achieve food security. The fundamental aim is to ensure the availability of good-quality food that promotes a healthy life for everyone. The achievement of this goal will require large access to food and the promotion of sustainable agriculture (United Nations Economic and Social Council (UNESCO),

2019). It will also require an increase in international investments and cooperation to bolster agricultural productive capacity in developing countries (United Nations Statistics Division (UNSD), 2016a, 2016b).

SDG-2 has eight targets to be achieved by 2030, as clearly stated in Target 2.1 of the SDG-2 of the United Nations Sustainable Development Goals: 'By 2030, end hunger and ensure access by all people, in particular the poor and people in vulnerable situations, including infants, to safe, nutritious and sufficient food all year round'. This means improving agricultural production, recognising and protecting small-scale farmers, and providing equal access to land, technology, sustainable food production systems, and resilient agricultural practices. Ending poverty and hunger is one of the top priorities in the 2030 Agenda for Sustainable Development. Achieving SDG-2 goals is the key for most developing countries, since it contributes to environmental sustainability and promotes economic growth. Food security and nutrition challenges are particularly critical in regions that are prone to multiple vulnerabilities and often cut across national boundaries, such as climate change hotspots. The nutrition of people in different horizons (areas) must be improved all year round. It is, therefore, important to reach the SDG-2 targets in these areas to improve agricultural sustainability and promote economic growth, environmental health, and social issues, such as improved standards of living (Barbier & Burgess, 2017). Furthermore, SDG-2 provides a global approach of achieving improved agricultural sustainability.

Several challenges have been raised in relation to achieving SDG-2 and the associated targets. For example, farmers in small-scale farming areas are earning less than the minimum wage. The land tenure system is also affecting production yields, and this have an impact on their efforts to meet targets (United Nations Economic and Social Council, 2019).

Many countries have revised their policies on wages, prescribing higher salaries for small-scale farmers (United Nations, 2015a). The land tenure system has also been revised in most countries, allowing access to land, inputs, and employment as a means to reduce poverty and malnutrition and to improve food security in the fight to end hunger by 2030 (United Nations, 2015a). All these efforts are to improve economic growth, which is fundamental to developmental issues in all countries.

Population growth has contributed significantly to food shortages, mostly in developing countries, due to the lack of resources (World Bank, 2017). Stress and degradation of natural resources have also been contributing factors of hunger. Resource conflict, mostly in rural areas of developing countries, has caused war, terrorism, and made people poorer, leading to malnutrition, deaths, and even migration of people to other countries (Nilsson, Griggs, & Visbeck, 2016).

On the other hand, the fight for zero hunger has promoted development in several rural areas, especially those in marginal areas where infrastructure development has been very poor and the lack of good schools, shops, healthcare centres, and even good water sources have been prominent. In addition, sustainable forestry, fisheries, and agricultural practices have also contributed in providing nutritious food for all and in generating a decent income, while also supporting rural development and protecting the environment. Food security can also be used to protect vulnerable women and children in any community.

The eradication of hunger across the world has seen targets and indicators aligned to four pillars of food security. These are availability, access, utilisation, and stability (Gil et al., 2019). Availability means having sufficient quality food at your disposal. The production of this food must also depend on a healthy environment. Access means having the physical capability to obtain a nutritious meal. Utilisation focusses on having an adequate dietary intake and the ability to absorb and use all the right nutrients in the human body. Finally, stability means ensuring that the other pillars are recognised and consistent on a daily basis. Achieving the sustainable development goals will automatically contribute to the eradication of all forms of hunger and malnutrition by 2030.

Ending hunger can have several positive impacts on the economy, health, education, and quality of goods produced by industry (Szabo et al., 2017; United Nations, 2019). The success of SDG-2 is connected to the success of other goals (Szabo, Mowlds, et al., 2016; Szabo, Nicholls, 2016; Szabo et al., 2017). A better-nourished population can have a greater positive impact on the economy (Szabo et al., 2017; United Nations, 2019).

According to the United Nations Economic and Social Council (2019), hunger is on the rise again globally, and undernourishment continues to affect millions of children after years of prolonged decline. Undernutrition reduces global gross domestic product (GDP) by up to USD 2 trillion per year (Gillespie & van den Bold, 2017). Africa remains the continent with the highest prevalence of undernourishment, affecting one-fifth of its population (more than 256 million people) (United Nations Economic and Social Council, 2019). Annual GDP losses due to malnutrition average 11% in Asia and Africa, which is greater than the loss experienced during the 2008–2010 financial crisis (Gillespie & van den Bold, 2017). In addition to undernutrition, the health landscape in all regions of the world is being drastically altered by an epidemic of another form of malnutrition: overweight obesity (Gillespie & van den Bold, 2017). Conflict, drought, and disasters linked to climate change are among the key factors causing this reversal in progress (FAO, IFAD, UNICEF, WFP & WHO, 2019).

It has also been reported that public investment in the agricultural sector is witnessing a decline globally. Government spending on agriculture compared to agriculture's contribution to the total economy has declined by 37%; the ratio fell from 0.42 in 2001 to 0.26 worldwide in 2017 (United Nations Economic and Social Council, 2019). This has prompted the need for support for small-scale food producers and an increase in investment for infrastructure and technology for sustainable agriculture (FAO et al., 2019; United Nations, 2015c).

Strengthening the resilience and adaptive capacity of small-scale and family farmers is critical to reversing the trend of the rise in hunger. This book will start by assessing and understanding the implications of SDG-2 for the economic, social, and environmental domains of communities and countries. In addition, three case study countries (Nigeria, Ghana, and The Gambia) will be used to record values of the UN indicators and illustrate the operationalisation of keeping track of these indicators. The book will also try to understand the connections between SDG-2 and the other SDGs in achieving the 2030 Agenda. Opportunities and challenges in implementing SDG-2 will be identified to foster progress and identify gaps to be considered during the implementation process. Finally, the book will present examples of best practice in implementing SDG-2 that can be upscaled or transferred to other regions and countries.

2

THE IMPORTANCE OF SDG-2

Sustainable development goal-2 (SDG-2) seeks to achieve food security and end or reduce hunger in all forms by 2030 (FAO et al., 2019). In addition, achieving SDG-2 indirectly contributes to the reduction of civil conflicts, migration, and security concerns. According to United Nations statistics, there are approximately 815 million hungry people around the world and 2 billion cases of malnutrition are expected by 2050. However, the zero hunger goal in the past quarters has also witnessed a decline in undernourishment from 19% to 11%, which contributed to the decline of stunting from 40% to 23% (United Nations, 2020a).

Ending hunger, achieving food security, improving nutrition, and promoting sustainable agriculture are fundamental to the developmental agenda of most countries in the world (FAO, 2016a; Gil et al., 2018). Agriculture is the single largest employer in the world and provides a livelihood to about 40% of the world's population (United Nations, 2020a). The sector still ranks as the largest source of income and employment for poor rural households, with 500 million small farms worldwide (most still rain-fed agriculture), providing up to

80% of the food consumed in a large part of the developing world (United Nations, 2020a).

The significance of this multi-dimensional goal can be subdivided into three interrelated components: ending hunger and improving nutrition (social dimension); achieving food security through productivity improvement and income increase (economic dimension); and promoting sustainable agriculture (environment dimension). With the above interrelation, it is obvious that achieving SDG-2 has economic, social, and environmental benefits and that sustainable agriculture has a key role to play in achieving this SDG. These benefits will now be summarised below.

2.1. ZERO HUNGER AND THE ECONOMIC BENEFITS

Zero hunger has a major impact on the economic growth of many countries. Ending hunger through improved and sustainable agriculture has a positive effect on human livelihoods, especially for poor, vulnerable populations, and small-scale farmers (Ballard, Kepple, & Cafiero, 2013; FAO, 2016b), seeing a significant number of women and men participating in the economic development of the country (IFPRI, 2015; United Nations, 2015c).

Achieving SDG-2 comes with an increase in access to knowledge, finance, and markets; it will strengthen agricultural production (Mollier, Seyler, Chotte, & Ringler, 2018). Infrastructure in rural areas will be improved, thus, increasing accessibility to rural communities for buyers, input suppliers, and resource distribution. Access to financial resources might attract more investment, thereby creating more employment opportunities for the local population (Mollier et al., 2018; Szabo et al., 2018), which favours the development of the rural economy (Mollier et al., 2018). This will result in rural communities having better access to clean water and improved health, shops, schools, and hospital facilities.

Agriculture is one of the major boosters of economic growth in many countries (Knox, Daccache, & Hess, 2013). Growth in agriculture is a key aid to the overall success of the economies of most countries. Improved agricultural productivity increases the supplies of raw materials for other industries such as the food, clothing, furniture, medicinal, and fuel industries (Gil et al., 2018; Mollier et al., 2018; United Nations, 2016). This might also lead to the emergence of non-agricultural industries, which may contribute to improving the availability of high-quality food (Mollier et al., 2018; United Nations, 2015b, 2016). The improvement of sustainable agriculture allows countries to produce more goods and services, which meet both domestic and export demands. An increase in agriculture sustainability production could be a good indicator to measure the growth of countries (Gil et al., 2018).

Zero hunger allows households to allocate scarce resources evenly and promote development to other income generation projects (Mollier et al., 2018). Households may allocate income to goods that support food security and reduce malnutrition; other goods not related to food security; and savings for future use (Leakey, 2017). Zero hunger improves micro- and macroeconomics allowing economic development. This may help countries to improve the balance of payments, attract investors leading to job creation, reduce poverty, bring in stability, maximise the use of resources, and bring in foreign currency (Gil et al., 2018; United Nations, 2016).

Achieving SDG-2 might promote the development of infrastructure through investments from new actors in the sector (Carlson et al., 2016; Zhang et al., 2015). These infrastructures can range from improved road networks, schools, hospitals, and business centres, important for the sale of farm products, to other indirect activities, for example, banking services, related to the processes of achieving SDG-2. Economic development in most countries is facilitated by the use of sustainable agriculture for food production. It is an important supplier of industrial raw materials such as trees for timber, maize for milling industries, herbal plants for medicines, soya beans for oil, and stock feed industries (Cabell & Oelofse, 2012). Such investment has led to the creation of well-paid jobs and promoted the development of communities, etc.

According to Gil et al. (2018), achieving SDG-2 is the basis for economic growth since it creates an opportunity for everyone to have access to healthy food and clean water. Thus, the basic commodities for life are made readily available for the entire population. The supply of these products will be higher than the quantities demanded; hence, business people would sell goods and services at reasonable prices. The availability of food resources in large quantities for all people may also promote fair competitive markets, allowing more people to participate (Mollier et al., 2018). Monopolies will also be very rare, especially for food-related goods due to food security and access to all.

Zero hunger promotes international trade and food market access, thereby allowing all to have access to the variety of food resources they need. This may also empower people economically and improve their income through the marketing of their products. Economic sustainability will be achieved due to zero hunger by 2030 if all countries in the world implement SDG-2 to achieve food security, improve agriculture, end hunger, and malnutrition with access to food for all (Gil et al., 2018; Mollier et al., 2018; United Nations, 2016).

2.2. ZERO HUNGER AND THE SOCIAL BENEFITS

Achieving zero hunger will have a positive impact on the societies where people live. Most societies are faced with challenges such as disease due to malnutrition, death due to food shortages, and migration resulting from the movement of people into other areas in search of food or other resources to improve food access. Countries where SDG-2 targets are implemented have shown great improvements in terms of food access, improved health, food security, for example, Brazil (Gil et al., 2018), and sanitation. These countries have also achieved human development, access to education, social development, and gender equality. Zero hunger has allowed people to send their children to school and promoted education (Szabo et al., 2018).

According to Barbier and Burgess (2017), quality education has seen a rise from 75% to 85% between the years 2000 and 2015 due to the adoption of zero hunger. Zero hunger may help countries to strengthen social behaviour through the reduction of crime, such as theft, murder, and even sexual abuse.

Educated societies have promoted political stability (Rosegrant & Hazell, 2000) and supported innovations that promote human development, such as new sustainable farming systems. Educated societies have also promoted societal development through the construction of toilets to reduce the spread of diseases such as cholera. Members of these societies are taught how to improve individual hygiene, domestic hygiene, and even community hygiene. Community hygiene is very important in society as it reduces neonatal death, reduces malnutrition, and promotes the proper disposal of any material, which may cause harm or danger to humans (FAO, 2016b; Mollier et al., 2018). Access to clean water is also achieved through community hygiene since people are taught how to obtain and preserve clean water. The achievement of zero hunger has also seen a decline in the spread of HIV/AIDS since hunger increases vulnerability to HIV/AIDS (Leakey, 2017) with more people being pushed into activities such as prostitution to make ends meet.

Zero hunger has the ability to achieve gender equality in a society. Programmes to promote the selection of women into positions of authority can be integrated into the overall agenda of SDG-2 when implemented at local, national, or regional levels. The education of young women will be promoted since there will be resources to pay for school fees and other needs. Statistics show that most societies facing hunger do not promote young women's education simply because there is the assumption that family affairs are run by men. Societies where zero hunger has been achieved have seen a rise in females competing with men for various positions in society, companies, and even politics. Target 2.3 of SDG-2 helps in increasing women's full and effective participation at all levels in the decision-making process (Gil et al., 2018).

Achieving zero hunger may help social communities move from civil unrest, war, and even political instability because most people have access to enough food, good nutrition, reduced poverty, good sanitation, and a green revolution (International Food Policy Research Institute, 2015; United Nations, 2016). In countries such as Somalia, civil unrest and war have been encouraged by food insecurity, with no access to resources leading to increased cases of malnutrition and death of children under the age of 5 years (Gil et al., 2018).

2.3. ZERO HUNGER AND THE ENVIRONMENTAL BENEFITS

The process of achieving zero hunger has a key role in the integrity of the natural environment. Improving agricultural sustainability leads to improved food security, which promotes environmental management (Gil et al., 2018). Sustainable agricultural systems promote soil erosion control measures that reduce soil infertility, water pollution, and the siltation of rivers, which are sources of water for animals and irrigation (Gil et al., 2018; Popkin, 2014). Control of soil erosion allows vegetation growth and improved soil moisture conservation (EU, 2017) which is beneficial to agriculture and the environment.

Sustainable agriculture is a key factor in afforestation and reforestation (RIVM, 2015), which are key drivers in reducing deforestation and desertification. Improved sustainable agriculture preserves soil quality through enhancing soil fertility (Assad et al., 2013), which improves environmental ecosystems and increases macro- and microorganism populations (Leakey, 2017). This may also improve nutrient cycling and energy flow in the ecosystem. Sustainable agriculture in the use of rotations and associations reduces the use of pesticides (EU, 2017), chemicals, and fertiliser that keep soil quality high, reduce water and soil pollution, and lower fresh water toxicity (EU, 2017; La Rovere, Pereira, Dubeux, & Wills, 2014).

The use of agroforestry systems in improving sustainable agriculture promotes soil quality improvement, restoration of soil quality, and protection of water-related ecosystems (Leakey, 2017). The technologies used to improve sustainable agriculture and food security may reduce the emissions of greenhouse gases, which have negative effects on ecosystems (Nepstad et al., 2014). This may promote a clean environment through the lowering of environmental temperatures due to afforestation, reforestation, and the use of agroforestry technologies. Countries where zero hunger has been achieved partially, for example, Brazil, have implemented environmental management policies and achieved high levels of environmental management. These polices have led to a reduction of water and land pollution and improved all types of ecosystem, be they aquatic, land-based, or terrestrial-based ecosystems (Gil et al., 2018).

Target 2.4 of SDG-2 ensures that sustainable food production systems and the implementation of sustainable agriculture are key to environmental benefits such as efficient nutrient cycling due to improved ecosystems resulting from improved agricultural production (Gil et al., 2018). Zero hunger highly facilitates land management as farmers try to reduce land degradation by using soil and water conservation methods such as infiltration pits, tied contours, and conservation tillage (Nyamadzawo, Gotosa, Moyo, & Govere, 2016).

Target 2.5 of SDG-2 maintains genetic diversity of seeds, cultivated plants, domesticated animals, and other animals kept on farms (Gil et al., 2018), which contributes to maintaining tree populations. Trees play an important role in environmental improvements as trees beautify the environment, ameliorate microclimates, and provide habitats for organisms (Leakey, 2017). Trees also play an important role in nutrient cycling due to deep roots, which absorb all leached nutrients and return them back to the surface. Tree leaves fall and provide mulching, which decomposes to release nutrients into the soil and increase plant growth, diversity, and increased food resources such as food, fuel, wood, and non-timber forest products (Nepstad et al., 2014).

There is no denying that the employment of sustainable agriculture as a strategy for combating zero hunger has a positive impact on the environment. The reduction or elimination of the use of fertilisers, agrochemicals, and pesticides on farming areas contributes to a reduced accumulation of chemicals in the soil which cause groundwater pollution, death of aquatic life, etc. (EU, 2017). In addition, the problem of expanding agricultural land so as to avoid the overuse of chemicals yet increasing agricultural output is reduced, thus, contributing to reduced deforestation and the adverse impacts on water-based ecosystems (FAO, 2014).

3

APPLYING SDG-2

The sustainable development goals (SDGs) are purposefully designed in a general manner with the intention for implementation to be specific and country led. In a more general sense, the SDGs offer guidelines, which are adaptable to specific contexts despite the need for flexibility rather than being prescriptive. However, the operationalisation and monitoring of the implementation requires tangible indicators and threshold values, which are important to ensure minimum standards and comparisons across countries. In the process of implementing the SDGs, emphasis is often shifted from the goals themselves by focussing on monitoring and assessing progress towards achieving the goals (Zinkernagel et al., 2018). Monitoring is key to informing priority actions and channelling scarce resources effectively (Gil et al., 2019).

Achieving SDG-2 will require the alignment of 'targets and indicators with the four pillars of food security: availability, access, utilisation and stability' (FAO, 2008). One of the main challenges when implementing SDGs is the selection of indicators that capture the interaction between the different pillars and overlapping areas with other SDGs (e.g. SDGs-2 and -3). Furthermore, recent research has highlighted a number of challenges encountered by countries when implementing the SDGs at a local level. Table 1 summarises some of the identified challenges.

According to Zinkernagel et al. (2018), the implementation of the SDGs at the local level could lead to the identification of synergies; it could also lead to contradictions in the achievement of the goal, as partly observed in Table 1. It might be important to conduct implementation at different levels with the use of different actors in accordance with the governing structure of each country, as well as a department responsible for monitoring progress. According to Zinkernagel et al. (2018), the challenge often lies in the identification of responsibility at each level, as well as establishing and operationalising cooperation. Legitimising an increasing number of indicators may help maintain the monitoring of these different responsibilities at a manageable level but may also be misused and lead to a situation in which areas with special interest in political debates are prioritised (Zinkernagel et al., 2018). In general, more knowledge and experience need to be gained over time to be in a position where an SDG could be used as a practical tool for achieving sustainable development. Furthermore, the development of effective policies to support the implementation of an SDG requires multiple iterations between theory and practice.

3.1. SDG-2 TARGETS AND INDICATORS

The SDG targets and indicators are intended to monitor progress towards the achievement of these goals. During the recently concluded 51st session of the UN Statistical Commission, the commission reviewed and approved multiple changes in the global indicator framework based on the 2020 Comprehensive Review conducted by the Inter-Agency

Table 1. Positive and Negative Effects of Each Identified Challenge Regarding the Implementation of SDGs at the Local Level (Adapted from Zinkernagel et al., 2018).

Identified Challenge	Negative	Positive
Access to reliable data	Difficult to gather data of sufficient quality for all indicators	More chance of being able to collect data for some indicators and opportunity to engage wider range of stakeholders
Policy relevance and political prioritisation	Risk that groups of indicators are ignored if defined as non-applicable for political reasons	Possibility to choose indicators that are policy-relevant at local level
Capturing the complexity of the implementation system	Despite the large number of indicators, interrelationships, interdependencies, and synergies are not highlighted or considered	The higher the number of indicators, the higher the likelihood of being able to capture complexity
Covering out of boundary challenges and externalities	Difficult for countries to generate data covering these elements	The higher the number of indicators, the greater the likelihood of covering out of boundary challenges and externalities
Acting in a multilevel and multi-actor governance system	Selection of a number of indicators that target levels other than the goal	Possibility of capturing issues that influence the goal but are dealt with at other levels
Availability of capacity and skills	The higher the number of indicators, the higher the need for skills in more areas of expertise, and the higher the need for capacity	Involving other competences and other actors will result in better cooperation and increased capacity across sectors

and Expert Group on SDG Indicators (IAEG-SDGs). This approval saw the revision of indicator 2.4.1 (*Revision of metadata: Change in 'use of biodiversity supporting practices' sub-indicator*) and indicator 2.5.2 (*Proportion of local breeds classified as being at risk of extinction*) and the addition of indicator 2.2.3 (*Prevalence of anaemia in women aged 15–49 years, by pregnancy status (percentage)*) for SDG-2.

Currently, SDG-2 is still composed of 8 targets and 14 indicators (see Table 2). The SDG 'targets are defined as aspirational and global, with each government setting its own national targets guided by the global level of ambition but taking into account national circumstances (UN General Assembly, 2015). Each Government will also decide how these aspirational and global targets should be incorporated into their national planning processes, policies, and strategies'. Table 2 shows the targets and indicators (previous and new) as supported by the United Nations SDG framework. The framing of the targets and indicators could lead to a variety of interpretations due to the vagueness of terms such as 'sustainable' or 'fair', as well as their lack of specificity regarding the scale of enforcement and monitoring, or the boundaries of 'food systems', and this exposes weaknesses in the original UN indicators (Gil et al., 2019).

In some countries, the list of indicators is not fully defined, nor are they able to report on the full list of indicators. SDG-2 indicators are based on input submitted by agencies and entities concerning their areas of expertise (Szabo et al., 2017). These agencies and entities are custodians of the indicators and support their monitoring. In the previous sets of indicators, nine were developed by the Food and Agriculture Organisation (FAO) of the UN, two were developed by the World Health Organisation, one was developed by the International Monetary Fund/World Bank, and the last two were developed by the Organisation of Economic Cooperation and Development.

Table 2. Previous and Updated Targets and Indicators (Adapted from United Nations Website: https://sustainabledevelopment.un.org/sdg2 and United Nations, 2020b).

Targets	Previous Set of Indicators	New Set of indicators
2.1 By 2030, end hunger and ensure access by all people, in particular, the poor and people in vulnerable situations, including infants, to safe, nutritious, and sufficient food all year round	2.1.1 Prevalence of undernourishment 2.1.2 Prevalence of moderate or severe food insecurity in the population, based on the Food Insecurity Experience Scale (FIES)	2.1.1 Prevalence of undernourishment 2.1.2 Prevalence of moderate or severe food insecurity in the population, based on the Food Insecurity Experience Scale (FIES)
2.2 By 2030, end all forms of malnutrition, including achieving, by 2025, the internationally agreed targets on stunting and wasting in children under 5 years of age, and address the nutritional needs of adolescent girls, pregnant and lactating women, and older persons	2.2.1 Prevalence of stunting (height for age <-2 standard deviation from the median of the World Health Organisation (WHO) Child Growth Standards) among children under 5 years of age 2.2.2 Prevalence of malnutrition (weight for height >+2 or <-2 standard deviation from the median of the WHO Child Growth Standards) among children under 5 years of age, by type (wasting and overweight)	2.2.1 Prevalence of stunting (height for age <-2 standard deviation from the median of the World Health Organisation (WHO) Child Growth Standards) among children under 5 years of age 2.2.2 Prevalence of malnutrition (weight for height >+2 or <-2 standard deviation from the median of the WHO Child Growth Standards) among children under 5 years of age, by type (wasting and overweight) 2.2.3 Prevalence of anaemia in women aged 15–49 years, by pregnancy status (percentage)

Table 2. (Continued)

Targets	Previous Set of Indicators	New Set of indicators
2.3 By 2030, double the agricultural productivity and incomes of small-scale food producers, in particular, women, indigenous peoples, family farmers, pastoralists and fishers, including through secure and equal access to land, other productive resources and inputs, knowledge, financial services, markets and opportunities for value addition, and non-farm employment	2.3.1 Volume of production per labour unit by classes of farming/pastoral/ forestry enterprise size 2.3.2 Average income of small-scale food producers, by sex, and indigenous status	2.3.1 Volume of production per labour unit by classes of farming/pastoral/forestry enterprise size 2.3.2 Average income of small-scale food producers, by sex and indigenous status
2.4 By 2030, ensure sustainable food production systems and implement resilient agricultural practices that increase productivity and production, that help maintain ecosystems, that strengthen capacity for adaptation to climate change, extreme weather, drought, flooding and other disasters and that progressively improve land and soil quality	2.4.1 Proportion of agricultural area under productive and sustainable agriculture	2.4.1 Proportion of agricultural area under productive and sustainable agriculture

2.5 By 2020, maintain the genetic diversity of seeds, cultivated plants, and farmed and domesticated animals and their related wild species, including through soundly managed and diversified seed and plant banks at the national, regional and international levels, and promote access to and fair and equitable sharing of benefits arising from the utilisation of genetic resources and associated traditional knowledge, as internationally agreed	2.5.1 Number of plant and animal genetic resources for food and agriculture secured in either medium or long-term conservation facilities 2.5.2 Proportion of local breeds classified as being at risk, not-at-risk, or at unknown level of risk of extinction	2.5.1 Number of plant and animal genetic resources for food and agriculture secured in either medium- or long-term conservation facilities 2.5.2 Proportion of local breeds classified as being at risk of extinction
2. a. Increase investment, including through enhanced international cooperation, in rural infrastructure, agricultural research and extension services, technology development, and plant and livestock gene banks to enhance agricultural productive capacity in developing countries, in particular least developed countries	2.a.1 The agriculture orientation index for government expenditures 2.a.2 Total official flows (official development assistance plus other official flows) to the agriculture sector	2.a.1 The agriculture orientation index for government expenditures 2.a.2 Total official flows (official development assistance plus other official flows) to the agriculture sector

Table 2. (Continued)

Targets	Previous Set of Indicators	New Set of indicators
2. b Correct and prevent trade restrictions and distortions in world agricultural markets, including through the parallel elimination of all forms of agricultural export subsidies and all export measures with equivalent effect, in accordance with the mandate of the Doha Development Round	2.b.1 Producer support estimate 2.b.2 Agricultural export subsidies	2.b.1 Agricultural export subsidies
2. c Adopt measures to ensure the proper functioning of food commodity markets and their derivatives and facilitate timely access to market information, including on food reserves, to help limit extreme food price volatility	2.c.1 Indicator of food price anomalies	2.c.1 Indicator of food price anomalies

Note: The global indicator framework was developed by the Inter-Agency and Expert Group on SDG Indicators (IAEG-SDGs) and agreed to as a practical starting point at the 47th session of the UN Statistical Commission held in March 2016 and updated during the 51st session of UN Statistical Commission held in March 2020.

At the same time, other organisations have put forward sets of indicators to monitor progress in SDG-2 (see Table 3). In most cases, the list of such indicators refer only to the goals of the SDG without bearing any relevance to a specific target in the 2030 Agenda. The proposal of new indicators as presented in Table 3 by Gil et al. (2019) was designed based on the availability of empirical data associated with monitoring progress. It is important to note here that even though the SDG targets are the same for all countries, the pathways used by countries to achieve the respective targets will be different based on the specificity of the local context, for example, data availability, local priority, local policy, etc.

According to the Gil et al. (2019), Targets 2.1–2.5 are directly related to food security and agricultural sustainability, whereas Targets 2a–2c are more market-related measures aimed at increasing agricultural investments and reducing market restriction, distortions, and volatility. Each of the targets is designed to achieve specific requirements of the SDG-2 as presented in the subsequent paragraphs using some targets.

SDG-2 Target 2.1 aims to 'end hunger and ensure access by all people, in particular the poor and people in vulnerable situations including infants, to safe nutritious and sufficient food all year round' by 2030 (World Bank, 2017). Largely, this can only be achieved through interactions with other SDGs such as 'no poverty'. The use of improved agricultural technologies and improved genetic resources, such as crop seeds and animals, may be helpful in ending hunger and improving nutrition for poor and vulnerable people. However, most people living in war zones are still failing to meet this target. These groups will, therefore, require the support of organisations such as the United Nations and other charity organisations (United Nations, 2015a).

Table 3. Review of the SDG-2 Targets and Indicators Proposed by the United Nations (Adapted from Gil et al., 2019).

SDG-2 Targets	Previous Set of Indicators (UN-IAEG-SDGs)	Revised List of Indicators by Gil et al. (2019)
2.1 By 2030, end hunger and ensure access by all people, in particular, the poor and people in vulnerable situations, including infants, to safe, nutritious, and sufficient food all year round	2.1.1 Prevalence of undernourishment 2.1.2 Prevalence of moderate or severe food insecurity in the population, based on the Food Insecurity Experience Scale (FIES)	2.1.1 Prevalence of undernourishment 2.1.2 Per capita food supply variability index 2.1.3 Depth of the food deficit
2.2 By 2030, end all forms of malnutrition, including achieving, by 2025, the internationally agreed targets on stunting and wasting in children under 5 years of age, and address the nutritional needs of adolescent girls, pregnant and lactating women and older persons	2.2.1 Prevalence of stunting (height for age <-2 standard deviation from the median of the World Health Organization (WHO) Child Growth Standards) among children under 5 years of age. 2.2.2 Prevalence of malnutrition (weight for height >+ 2 or <– 2 standard deviation from the median of the WHO Child Growth Standards) among children under 5, disaggregated by type (wasting and overweight)	2.2.1 Prevalence of stunting among children under 5 years of age 2.2.2.a Prevalence of wasting among children under 5 2.2.2.b Prevalence of underweight among children under 5 2.2.3 Prevalence of anaemia among pregnant women 2.2.4 Average protein supply 2.2.5 Share of average protein supply of animal origin 2.2.6 Prevalence of obesity among adults

2.3 By 2030, double the agricultural productivity and incomes of small-scale food producers, in particular women, indigenous peoples, family farmers, pastoralists and fishers, including through secure and equal access to land, other productive resources and inputs, knowledge, financial services, markets and opportunities for value addition and non-farm employment	2.3.1 Volume of production per labour unit by classes of farming/pastoral/ forestry enterprise size. 2.3.2 Average income of small-scale food producers, by sex and indigenous status	2.3.1 Yield gap 2.3.2 Rural poverty headcount ratio at national poverty lines 2.3.3 Prevalence of farmers earning less than the national minimum wage
2.4 By 2030, ensure sustainable food production systems and implement resilient agricultural practices that increase productivity and production that help maintain ecosystems, strengthen capacity for adaptation to climate change, extreme weather, drought, flooding and other disasters, and progressively improve land and soil quality	2.4.1 Percentage of agricultural area under sustainable agricultural practices 2.4.2 Percentage of agricultural households using irrigation systems compared to all agricultural households 2.4.3 Percentage of agricultural households using eco-friendly fertilisers compared to all agricultural households using fertilisers	2.4.1 Water withdrawn by agriculture as a percentage of total withdrawal 2.4.2 Average water productivity in agriculture 2.4.3 Nitrogen use efficiency 2.4.4 Average nitrogen surplus 2.4.5 GHG emission intensity of food production 2.4.6 Average carbon content in the topsoil 2.4.7 Climate change vulnerability index for food 2.4.8 Use of pesticides per area

Table 3. ***(Continued)***

SDG-2 Targets	Previous Set of Indicators (UN-IAEG-SDGs)	Revised List of Indicators by Gil et al. (2019)
2.5 By 2020, maintain the genetic diversity of seeds, cultivated plants, and farmed and domesticated animals and their related wild species, including through soundly managed and diversified seed and plant banks at the national, regional and international levels, and promote access to and fair and equitable sharing of benefits arising from the utilisation of genetic resources and associated traditional knowledge, as internationally agreed	2.5.1 Number of plant and animal genetic resources for food and agriculture secured in either medium or long-term conservation facilities 2.5.2 Proportion of local breeds classified as being at risk, not-at-risk, or at unknown level of risk of extinction	2.5.1 Average number of gaps in ex situ collections of selected crop gene pools 2.5.2 Proportion of local breeds classified as being at risk out of all breeds whose risk of extinction is known

Another target (Target 2.2) of SDG-2 is to end all malnutrition by 2030 across the world (United Nations, 2015c). The burden of malnutrition is recognised as one of the triple burdens that have crippled and troubled countries around the world. These burdens are the co-existence of undernourishment, micronutrient deficiency, and over nutrition, which is manifested as being overweight and obese. Ending malnutrition will require proper planning and monitoring strategies to improve agricultural production across the world. This target also needs to end malnutrition for pregnant and lactating women, adolescent girls, and older persons who are not able to grow their own food. To end malnutrition, there is a need to provide food with high levels of protein since most people affected by malnutrition lack protein in their diets.

However, the measurement of complex nutrition levels is also not being carried out in developing countries. There is a lack of staff and expertise in the many departments in the health sector in developing countries. The main challenges have been to provide the same proportional efforts being given to developed countries. Over time, in many developing countries, there has been a lack of monitoring of vulnerable populations and equitable access to food. This has led to many food programmes being politicised with food being distributed according to political party lines (United Nations, 2019) rather than to the poor and vulnerable populations. Agriculture's share of government expenditure increasingly lags behind its economic contribution. According to the United Nations Statistics Division (UNSD) (2016a, 2016b), 'the productive capacity of agriculture depends on investments from public and private, domestic and foreign sources'. Recent trends in government spending have not been favourable. The Agriculture Orientation Index (AOI) – the agriculture share of government expenditure divided by the agriculture share of GDP – fell from 0.37 to 0.33 between 2001 and 2013 in developing countries (UNSD, 2016). The decline was interrupted only during the food price crisis of 2006–2008, when governments boosted agricultural spending (UNSD, 2016a, 2016b).

Ending malnutrition can be achieved by meeting Target 2.3, which is to double agricultural productivity and income for small-scale farmers (Barbier, 2016). These people are mostly affected by hunger because they are resource poor and due to the yield gap between small-scale farmers found in most developing countries. Poverty in small-scale farming areas is a major challenge to meeting Target 3, but by interacting SDG-2 with SDG-1, the target can be easily met. In most continents, countries are creating measures and means to end poverty. Indirectly, these measures are contributing to ending hunger, improving sustainable agriculture, and improving food security (FAO, 2016a; United Nations, 2015a).

SDG-2 Target 2.5 is aimed at the conservation of agro-biodiversity, that is, the diversity of living organisms used in agriculture, to increase food quality and health. This helps in the achievement of sustainable agriculture and improves nutrition. Conservation of agro-biodiversity has been noted to increase soil fertility, which increases plant growth leading to improved food productivity. The use of the soil and water management practices (rainwater harvesting techniques) of tied contours and infiltration pits has been recognised as increasing agricultural productivity (Nyamangara et al., 2013) and improve environmental aspects and economic growth.

The original UN indicators have revealed two weaknesses, which include (1) the targets and indicators that do not always focus on the same groups of people, and (2) although indicators were phrased quantitatively, unclear concepts hinder their quantification (Gil et al., 2018). For example, indicator 2.4.1 refers to the percentage of agricultural land that is under sustainable practices. While the percentage is a quantifiable metric, agreeing on what sustainability is, when it is achieved and what it translates into at different scales can be difficult (Gil et al., 2018). It is important to note that the proposition of meaningful indicators and the monitoring of progress in countries are highly determined by the available information and their quality. This is paramount to investing in better global databases (Alkire & Samman, 2014).

Further examples, according to Gil et al. (2018) to demonstrate flaws within the existing indicators, show that Targets 2.3 (agricultural productivity), 2.4 (sustainability of food production systems), and 2.5 (genetic diversity) are less clearly defined and not always universally relevant. While the main challenge for Targets 2.1 (hunger) and 2.2 (malnutrition) are more on how the targets could be achieved efficiently, Targets 2.3–2.5 require first a definition before verifying the operationalisation of the targets.

It is important to note that not all targets have the same degree of priority in different countries, as is illustrated with the three countries assessed in Section 3.2. Agricultural practices based on the specificities of each country and their priority areas should be fostered, especially those practices that contribute to the achievement of multiple sustainability targets at the same time. In addition, as a result of the aspirational nature of the SDGs, setting national thresholds and realistic policy targets are necessary steps towards their implementation. 'The SDG-2 indicators should be prescriptive enough to ensure comparability across countries and the adoption of minimal standards worldwide and flexible enough to account for country's specific challenges and demands' (Gil et al., 2019).

In addition, before starting the process of implementation, it is important to conduct a more reliable assessment of global figures and policy while complementing them with country-level data whenever possible. The formulation of targets and indicators for SDG-2 should adopt a minimal standard worldwide so as to ensure comparability across countries and be flexible enough to account for the specific challenges and demands (Gil et al., 2018). It is important to note that great diversity might exist also at the sub-national level, which makes it necessary to get as much information as possible. Therefore, reaching a perfect balance requires a participatory approach and should take advantage of existing sustainability frameworks (Slätmo, Fischer, & Röös, 2016). Often, several iterations between theory and practice are required for the development of an effective policy.

Achieving SDG-2 might seem to be ambitious, as it requires all countries to be developed, peaceful, and to recognise human rights. Many people around the world rely on agriculture as their main source of living; however, they still struggle to access productive resources and are highly vulnerable to the climate and other disturbances. Therefore, with these challenges, achieving the 'zero hunger' vision under Agenda 2030 requires social protection and food systems that are economically efficient, socially inclusive, and environmentally sustainable. Ensuring that there is sufficient food for all requires widespread promotion of sustainable agriculture and conservation of biodiversity; enhanced livelihoods of smallholder farmers, including fishers, foresters, and pastoralists; and increased investments in agriculture, markets, agribusinesses, and related infrastructures.

Cooperatives are tested models that can sustainably support small-scale producers and family farmers in achieving economies of scale and improving their access to markets, financing, information, and other resources. According to the

International Labour Organisation (ILO) (2018), cooperatives can be broadly classified into two main groups, which include agricultural producers' cooperatives and consumers' cooperatives. While the agricultural producers' cooperatives help farmers to increase their productivity and incomes, the consumers' cooperatives contribute to food security and sustainable agriculture at the end of the food supply chains. In addition, there are an increasing number of new initiatives that promote sustainable and fair food production (ILO, 2018). A good example of such an initiative is Community Supported Agriculture (CSA). Here, the consumers established a long-term and direct partnership with farmers. A major outcome of such a partnership is to share the costs and benefits of products from the farm. By 2015, about 6,300 CSAs were in operation in Europe, producing food for a million consumers in a sustainable manner.

3.2. TRACKING SDG-2

It is sometimes hard to measure the exact contribution of interventions towards the implementation of the SDG-2 targets or the targets of any other SDG. Often, indicators are used to keep track of the progress made in the implementation of specific SDGs. While some of the indicators used for monitoring SDG-2 present objectives, which could be straightforward objectives or complex with conditional actions, others present the means to achieve the objectives (Gil et al., 2019). Currently, there are 14 indicators, as presented in Table 2, used for tracking progress in the implementation of SDG-2. One of the major inconsistencies still faced in the use of indicators is the lack of data to monitor progress and the operationalisation of indicators that are still very unclear (Hák, Janoušková, & Moldan, 2016).

To address the issue of historical inconsistencies in SDG data, the UN Inter-Agency Expert Group developed a tier system. In addition, the tier system is intended to assist in the development of global implementation strategies. According to the United Nations (2020b), the tier system is classified under the following criteria:

'Tier 1: Indicator is conceptually clear, has an internationally established methodology, and standards are available, and data are regularly produced by countries for at least 50% of countries and of the population in every region where the indicator is relevant.

Tier 2: Indicator is conceptually clear, has an internationally established methodology, and standards are available, but data are not regularly produced by countries.

Tier 3: No internationally established methodology or standards are yet available for the indicator, but methodology/standards are being (or will be) developed or tested. (*As of the 51st session of the UN Statistical Commission, the global indicator framework does not contain any Tier III indicators.*)

In accordance with the tier system framework, indicators are designed based on the availability of data. In other words, SDG indicators should, by definition, make use of methodologies that are in use already. In other cases, authors such as Gil et al. (2019) have reviewed the UN-proposed SDG-2 indicators and suggested an improved list as presented in Table 3. In the following section, agriculture and food security in Nigeria, Ghana, and The Gambia will be analysed using UN SDG-2 indicators to present the various food-related challenges as well as to illustrate the operationalisation of SDG-2 across different development contexts. The section starts by briefly presenting the countries priorities and policies in achieving the SDG-2.

3.3. CASE STUDIES – COUNTRY PRIORITIES AND POLICIES

Three countries (Nigeria, Ghana, and The Gambia) from the African continent have been selected randomly to illustrate the operationalisation of SDG-2 at a country level. Table 4 presents an overview of macro-economic factors and the agro-food sector of the three countries.

Table 4. Overview of the Three Selected Countries.

	Nigeria	Ghana	The Gambia
Macro-economic factors			
Total area (km^2)	923, 768	238, 535	11, 295
Population (2019)	200 million	30 million	2 million
GDP per capita (US$, 2019)	2,222 (138th)	2,223 (137th)	755 (171st)
HDI (0–1) (2018)	0.534 (158th)	0.596 (142nd)	0.466 (174th)
Agri-food sector			
Employment share (%)	70%[a]	52% (FAO, 2020a)	70% (FAO, 2020b)
GDP share (%)	20%[a]	54% (FAO, 2020a)	30% (FAO, 2020b)
Export share earnings (%)	15% (Nwachukwu, 2014)	40% (FAO, 2020a)	40% (FAO, 2020b)
Predominant farm size	< 2 ha	< 1.6 ha	< 2 ha
Prevalent farming characteristics	Predominantly small-holder, traditional and rain-fed	Predominantly small-holder, traditional and rain-fed	Predominantly small-holder, traditional and rain-fed

HDI, Human Development Index.
[a]Synthesis Report of the Nigeria Zero Hunger Strategic Review, 60pp.

3.3.1. Nigeria

According to the *Synthesis Report of the Nigeria Zero Hunger Strategic Review*, Nigeria is faced with chronic food insecurity, as its human and economic costs are immense. The Nigerian *Agricultural Promotion Policy – 2016–2020* focusses on ensuring food security, institutional reforms, and incentives to technological development at the local level. The current enabling environment for agriculture is provided by the *Green Alternative: The Agriculture Promotion Policy* (APP: 2016–2020), which aims at resolving major bottlenecks in food production and marketing. The objectives of the *Green Alternative: The Agriculture Promotion Policy* are complementary and mutually reinforce the Zero Hunger Strategic Review of Nigeria.

Small-holder agricultural producers constitute the main producers of food for the urban population. The small-holder producers are mostly rural dwellers and depend on agriculture for their livelihood. The vast majority of small-holder farms often fail to produce surpluses for small- and medium-scale enterprises across rural Nigeria. In addition, Nigeria is often subject to periodic droughts and floods, which have an adverse impact on the agricultural output, thus, increasing the vulnerability of populations, especially in rural areas. It is, therefore, important to strengthen small-holder-farming systems, since they will contribute to a higher output and support the growth of the national economy (Synthesis Report of the Nigeria Zero Hunger Strategic Review, 60pp).

The implementation of the Agricultural Transformation Agenda (2010–2014) saw a rise in national food production by 21 million tonnes, food imports significantly decline, and the number of farm jobs rises by 3.6 million (Federal Ministry of Agriculture and Rural Development (FMARD), 2012). One of the major challenges facing agricultural legislation is the slow pace of passing bills into law and the inadequate compliance by executive authorities (Synthesis Report of the Nigeria Zero Hunger Strategic Review, 60 pp).

3.3.2. Ghana

The agricultural sector provides over 90% of the food consumed in Ghana with approximately 39% of the farm labour force being women. Thus, a larger number of the population derive their livelihood directly or indirectly from the agricultural sector with it also serving as the main source of poverty reduction. Cocoa is the most important agricultural export crop, normally accounting for 30–40% of total exports.

The agricultural practice is predominantly small-holder, traditional, and rain-fed, which varies with agro-ecological zones in the country. For example, in the forest zone, tree crops are significant with cocoa, palm oil, coffee, and rubber being of particular importance. The food crops in this area are mainly inter-cropped mixtures of maize, plantain, cocoyam, and cassava. However, rice remains important in all the zones.

There is no lack of policies, plans, strategies, programmes, and projects designed to achieve food and nutrition security. Since 1992, there have been five national development plans and food security has been an important goal in all the plans (Steiner-Asiedu, Dittoh, Newton, & Akotia, 2017). The main problem has been a lack of implementation, which is also coupled with shortcomings from some stakeholders saying that there is hardly any relationship between the plans and the actual situation on the ground. In addition, another weakness of the planning and policies has been inadequate understanding and a lack of sensitisation and engagement with stakeholders by those who are expected to implement the plans, policies, etc., at regional and district levels (Steiner-Asiedu et al., 2017).

3.3.3. The Gambia

The Gambia is the smallest country in mainland Africa. One of the main drivers of growth in the Gambia is agriculture but it remains very vulnerable to external shocks. For example, the drought in 2011 reduced agricultural production by 24.2% from the previous year (Gambia Bureau of Statistics (GBOS), 2016). The farming economy is the only means of income creation for the majority of the rural population. Under the Millennium Development Goals (MDGs) framework, The Gambia took great strides in reducing infant mortality rates, maternal mortality rates, improving access to rural water supplies, and achieving gender parity in primary and secondary education (Gambian Ministry of Finance and Economic Affairs, 2014). Based on this progress, The Gambia has further committed to closing the remaining gaps under the Sustainable Development Goals (SDGs), which include amongst other targets SDG-2.

The agricultural sector comprises different sub-sectors, including field crops, horticulture, livestock, forestry, and fisheries. Crop production is very diverse. The predominant mode of production is characterised by traditional methods of farming that depend on rainfall and have inadequate access to financial and agricultural resources. In addition, female farmers tend to be more vulnerable, as they have limited access to land and productive capital, which reduces their ability to secure financial services.

The agriculture sector is challenged by poor infrastructure, soil fertility depletion, declining agriculture commodity prices, soaring prices of production inputs, and low private investment. The country continues to depend on imports of many agricultural products, particularly refined sugar and linseed oil, both of which constitute 26% of the total import value of agricultural products (FAO et al., 2018). However, The

Gambia is limited in importing food because of a problematic balance of payments, which restricts their borrowing and purchasing capacity of food in the international market (FAO, 2016d).

The Gambia has existing policies that are relevant to SDG-2 targets. Through the National Development Plan (NDP) 2018–2021, the government has committed to strengthening support for the implementation of SDG-2. There are also several ongoing public sector projects that are being funded by various organisations aimed at reducing poverty, improving food and nutrition security, cultivating productivity, and strengthening resilience.

Although The Gambia is one of the countries least responsible for climate change, their highest emissions (79%) come from the agricultural sector with about 65% coming from the livestock sub-sector where enteric fermentation contributes to about 36% of methane emissions, while manure left on pastures constitutes about 26% of emissions from methane and nitrous oxide (FAO, 2016c).

3.4 APPLICATION OF THE SDG-2 TARGETS AND INDICATORS

Table 5 shows measurements of the respective SDG-2 indicators in Nigeria, Ghana, and The Gambia by accessing the Global SDG Indicators database of the United Nations Statistics Division and the database of WHO. Experience shows that many of the datasets for indicators are not available for the chosen countries. It is important to note that the Global SDG Indicators database is a platform that provides access to data compiled through the UN System in the preparation for the Secretary-General's annual report on *Progress towards the Sustainable Development Goals*. This strongly recommends

Table 5. Application of SDG-2 Targets and Indicators in Nigeria, Ghana, and the Gambia.

Indicator	Updated Tier Classification (by IAEG-SDG Members)	Data sources	Unit	Ref Year	Nigeria	Ghana	The Gambia
2.1.1 Prevalence of undernourishment	I	United Nations Statistics Division	%	2015	9.6	6.2	10.2
				2017	13.4	5.5	10.2
2.1.2 Prevalence of moderate or severe food insecurity in the population, based on the Food Insecurity Experience Scale (FIES)	I	United Nations Statistics Division	%	2015	36.4	49.55	NA
				2017	36.43	49.55	54.06
2.2.1 Prevalence of stunting (height for age <-2 standard deviation from the median of the World Health Organization (WHO) Child Growth Standards) among	I	United Nations Statistics Division	% of children aged 0–59 month	2013	36.48	NA	24.56
				2018	36.81	NA	13.48
2.2.2 Prevalence of malnutrition (weight for height >+2 or <-2 standard deviation from the median of the WHO Child Growth Standards) among children under 5 years of age, by type (wasting and overweight)	I	United Nations Statistics Division	overweight (%)	2013	5.1	NA	3.2
				2018	2.1	NA	2.5
			wasting (%)	2013	18.05	NA	11.01
				2018	6.70	NA	6.03
2.2.3 Prevalence of anaemia in women aged 15 to 49 years, by pregnancy status (percentage)	Pending data availability review	World Health Organization (WHO)	%	2013	58.71	56.1	62.12
				2016	57.8	54.25	61.81

2.3.1 Volume of production per labour unit by classes of farming/pastoral/forestry enterprise size	II	United Nations Statistics Division	Constant USD	2013	16.29	NA	NA
				2016	11.92	NA	NA
2.3.2 Average income of small-scale food producers, by sex and indigenous status	II	United Nations Statistics Division	Constant USD	2013	789.01	1044.76	NA
				2016	921.54	NA	NA
2.4.1 Proportion of agricultural area under productive and sustainable agriculture	II	United Nations Statistics Division	(%)	2015	NA	NA	NA
				2017	NA	NA	NA
2.5.1 Number of plant and animal genetic resources for food and agri-culture secured in either medium- or longterm conservation facilities	I	United Nations Statistics Division	Number	2015	0	0	0
				2019	0	0	0
2.5.2 Proportion of local breeds classified as being at risk of extinction	II	United Nations Statistics Division	%	2015	NA	NA	NA
				2019	NA	NA	NA
2.a.1 The agriculture orientation index for government expenditures	I	United Nations Statistics Division	Index	2004	0.12	NA	NA
				2013	0.09	NA	NA
2.a.2 Total official flows (official development assistance plus other official flows) to the agriculture sector	I	United Nations Statistics Division	Constant USD	2015	154.46516	167.99729	7.72789
				2017	309.60984	143.16432	11.03872

Table 5. (Continued)

Indicator	Updated Tier Classification (by IAEG-SDG Members)	Data sources	Unit	Ref Year	Nigeria	Ghana	The Gambia
2.b.1 Agricultural export subsidies	I	United Nations Statistics Division	Millions of current United States dollar	2015	NA	NA	NA
				2018	NA	NA	NA
2.c.I Indicator of food price anomalies	II	United Nations Statistics Division	Index	2016	Maize = 1.8 millet = 2.0 Rice = 17.0 Sorghum = 2.0 Wheat = NA	Maize = – 0.5 Millet = 0.9 Rice = 0.6 Sorghum = –1.1 Wheat = NA	Maize = NA Millet = NA Rice = NA Sorghum = NA Wheat = NA
				2017	Maize = –0.1 Millet = 0.3 Rice = – 0.3 Sorghum = 0.2 Wheat = NA	Maize = – 0.9 millet = – 0.6 Rice = – 0.5 Sorghum = – 0.7 Wheat = NA	Maize = NA Millet = NA Rice = NA Sorghum = NA Wheat = NA

the update of the database so that a proper account of progress can be reported.

Despite the lack of data for some indicators in Table 5, the results for the respective countries could highlight the possible differences in their priority areas. Ghana represents the country with the highest missing values for their indicators. Overall, The Gambia performed poorly in indicators 2.1.2 and 2.2.3 even though a decreasing trend is observed in the latter. This leaves no doubt that there is a need for improvement in relation to these indicators. Nigeria witnessed an increase with indicator 2.1.1 from 9.6% to 13% in the prevalence of undernourishment. Nonetheless, most of the other indicators either saw a decreasing trend or remained the same, which for the most part is still significantly high and requires improvement. Other points for improvement are indicator 2.3.2 for Nigeria and indicator 2.a.2 for Nigeria and The Gambia, even though they experienced an increasing trend, which was positive. Note that the results presented here do not say much on whether the countries are on the right track to achieving SDG-2 targets since they need to be observed over a longer period to identify trends.

The monitoring of countries' progress is conditional on information availability and quality. Analysis shows that it is paramount to invest in a better global database. When implementing the SDGs, it is important to check the reliability of existing global figures and complement them with country-level data whenever possible. Furthermore, data limitations elicit the need for a more holistic policy approach when implementing the SDG agenda.

4

UNDERSTANDING THE INTERLINKAGES BETWEEN SDG-2 AND THE OTHER SDGS

Sustainable development goal-2 (SDG-2) is the key to the success of the entire SDG agenda because of its intrinsic relationship to society, the economy, and the environment (FAO, 2016a). By design, all SDGs interact with one another because they are a set of integrated global priorities and objectives that are fundamentally interdependent. The targets of the respective SDGs are not exempted from such interaction. According to Nilsson et al. (2016), there could be positive and negative interactions between SDG goals. For example, positive interaction could range from enabling, reinforcing, and even indivisibility of goals, whereas negative interaction could range from goals being constrained, counteracted or, in the worst case, cancelling one another out.

According to the United Nations Division for Sustainable Development (UN-DESA) (2018), being aware of the interlinkages between the SDGs is critical for policy coherence towards sustainable development. In addition, being aware of the interlinkages between SDGs helps to prioritise action with larger synergetic impact and facilitates policy dialogue and learning, which prompts effective investments. This is crucial for achieving long-lasting sustainable development outcomes. Table 6 briefly describes an overview of the interactions between SDG-2 and the other 16 SDGs. The interactions presented in Table 6 are largely summarised from Mollier, Seyler, Chotte, and Ringler (2017); they show the context dependency of the interactions and provide a more practical characterisation of SDG-2 interactions among the other 16 'integrated and indivisible' SDGs.

Table 6. Interdependencies of SDG-2 with other SDGs.

Interacting SDGs	Characterisation of Interaction Between the SDGs
Goal 2 + Goal 1	'Eradicating poverty cannot be achieved without ensuring food and nutrition security for all'
Goal 2 + Goal 3	'Health and well-being cannot be achieved without access to a sufficient quantity and quality of food'. 'SDG-3 supports SDG- 2, because a healthy population is essential for achieving nutrition and agricultural production targets'
Goal 2 + Goal 4	'SDG 4 on ensuring quality education enhances people's ability to adapt their nutrition, hygiene, sanitation and health practices and to demand appropriate services'
Goal 2 + Goal 5	'Achieving the targets related to access to food, quality nutrition for all, and agricultural incomes will provide key enabling conditions for women's empowerment and gender equality as it opens up development opportunities for women'
Goal 2 + Goal 6	'Food production is strongly dependent on and affects the quality and availability of water, because boosting agricultural production can increase water withdrawals and worsen land and water degradation'
Goal 2 + Goal 7	'Agriculture, food production, and consumption are strongly dependent on energy services; conversely, biomass, and agricultural waste are potential sources of renewable energy'

Table 6. (Continued)

Interacting SDGs	Characterisation of Interaction Between the SDGs
Goal 2 + Goal 8	'Increasing sustainable agricultural productivity and incomes of small-holder women and men, SDG-2 can participate in sustainable economic growth'. 'Especially in remote rural areas that are cut off from most alternative employment opportunities, agriculture is often the only viable source of both employment and food and nutrition security'
Goal 2 + Goal 9	'Developing and upgrading rural infrastructure, integrating small-scale enterprises into value chains and enhancing investment in agricultural research are aligned with SDG-9'. 'Such infrastructure may also accelerate biodiversity loss, over extract of water resources, and ignite other unsustainable practices'
Goal 2 + Goal 10	'Empowering small-scale food producers, both women and men (who represent an important segment of the world's extreme poor) and ensuring their equal access to resources such as land, facilitates the reduction of inequality'
Goal 2 + Goal 11	'Urban agriculture, thus, can contribute to social welfare and sustainable development of cities and can support development of green spaces. It can also contribute to waste avoidance and recycling of organic waste in cities' (Goldstein et al., 2016)
Goal 2 + Goal 12	'The subsequent efficiency, waste, and loss reduction targets and the aim to manage chemicals more judiciously directly support SDG-2 in terms of increased productivity and more sustainable natural resource use'
Goal 2 + Goal 13	'Sustainable agricultural practices play an important role in climate adaptation and mitigation (such as improving soils and land quality, genetic diversity, and bioenergy)'
Goal 2 + Goal 14	'More sustainable ocean fisheries and better access for small-scale fishers and residents of small-island states to these resources will support food security and nutrition in the long term'

Table 6. (Continued)

Interacting SDGs	Characterisation of Interaction Between the SDGs
Goal 2 + Goal 15	'Sustainable agricultural systems and practices contribute to ecosystem health. However, increased agricultural production and productivity, if not sustainable, can result in deforestation and land degradation, jeopardising long-term food security'
Goal 2 + Goal 16	'Hunger and food insecurity are sources of political instability, conflict and war – to the point that hunger is, at times, deliberately used in conflicts as a weapon to starve opponents into submission (seizing or destroying food stocks, livestock, cutting off marketed supplies of food, targeting farmers, land-mining, etc.)'
Goal 2 + Goal 17	'Enhancing multi-stakeholder partnerships, data, monitoring and accountability, and especially policy and institutional coherence, should also positively impact SDG'

Table 6 demonstrates that while most of the goals are relevant for achieving SDG-2, two, in particular, have a broader and cross-cutting relevance to the achievement of Agenda 2030. Specifically, SDG 16 on promoting peace, justice, and strong institutions, and SDG 17 on partnerships for goals, can both be conceived more as enablers rather than just goals in themselves, since none of the other SDGs will be achieved without them. SDG-2 encourages initiatives that provide opportunities for strategic and operational partnerships that enhance prospects for progress among all the other SDGs. Achieving SDG-2 is, thus, likely to invoke multiple synergies and trade-offs with other SDGs, across temporal and spatial scales, ultimately underscoring the indivisible nature of the SDG agenda (Gil et al., 2019). SDGs must, therefore, be considered as a single development agenda that calls for a comprehensive and integrated policy framework (Le Blanc, 2015; von Stechow et al., 2016).

Policy integration and coherence are a necessary condition to successfully implement the 2030 Agenda. The successful implementation of SDGs in countries will require a change in procedures and institutions because of the sometimes dense and different types of interlinkages between the SDGs. In such contexts, an in-depth understanding of the interlinkage between SDG 2 and the other SDGs for that specific country is critical to provide specific management options with minimum trade-offs. Therefore, a multi-sector approach that considers the environmental boundaries of governmental departments and all other stakeholders is very important for policy integration and coherence. Policy coherence for sustainable development is enshrined in Target 2.2.

Often countries do not classify the interlinkages between SDGs; however, they often acknowledge the need to adapt and change their procedures and institutions to implement them. According to Tosun and Leininger (2017), there was no systematic relationship between the assessment of interlinkages by the implementing countries and their approaches for procedural and institutional change. This has a serious implication in policy making or implementation of the SDGs, since there is the likelihood that such an approach will neglect the need for the integration of the relationships between the types of interlinkages and the types of procedural and institutional reforms. For many if not all goals, having in place effective governance systems, institutions, partnerships, and intellectual and financial resources is key to an effective, efficient, and coherent approach to implementation (Mollier et al., 2017).

There is no scientific evidence for whether policy coherence will really yield better policy results (Tosun & Leininger, 2017). The SDGs were adopted in 2015 and this provides a good base for assessing the level of policy integration within countries. Future research should perhaps study how various intergovernmental organisations position themselves to attain policy coherence in their overall strategy for successful implementation of the SDGs or their respective thematic areas. One key importance of such a study is that it contributes to stimulating a more science-policy dialogue on the interaction between the SDGs. According to the International Council for Science (2017), this approach provides a starting point for policy-makers and other stakeholders to set their priorities and implementation strategies, and to engage the policy community in further developments of knowledge in this field. A more scientifically informed analysis of the interactions between the SDGs would support more coherent and effective decision-making and facilitate better follow-up and monitoring of progress (International Council for Science, 2017).

5

ACHIEVING SDG-2 – OPPORTUNITIES AND CHALLENGES

Sustainable development goal-2 (SDG-2) provides opportunities and challenges at different levels. Furthermore, the SDG-2 targets have multiple reinforcing and constraining linkages with the other 16 SDGs that also provide both challenges and substantial scope for solutions (ICSU, 2017). This is true because in Africa, most governments see the implementation of the SDGs as an opportunity to foster African development. According to the former African Union Commission Chairperson Nkosazana Dlamini-Zuma, 'the post 2015 development agenda is a critical milestone towards the realisation of agenda 2063 and this provides a good opportunity for Africa to finish the implementation of the incomplete United Nation Millennium Development Goals (MDGs)'. The SDGs mobilise global action towards clearly defined goals and targets within a time frame from 2015 to 2030. 'The translation of SDGs into regional development strategies not only enhances the link between global objectives and the multi-faceted reality of a multipolar world, but more importantly provides a strategic option to strengthen the bargaining power of African countries' (Valensisi & Karingi, 2017).

The strong interlinkage between SDG-2 and related goals and targets presents challenges that can only be addressed on a case by case basis, considering that the situation changes from country to country. Governments are addressing these challenges through the establishment of national platforms with integrated policies and programmes (High-level Political Forum on Sustainable Development (HLPF), 2017). There is wide consensus that the international community should continue to encourage these efforts and provide space to exchange knowledge and foster productive partnerships. There have been encouraging policy developments at the global level relating to SDG-2 in recent years. The transformational change needed to realise the 'zero hunger' vision of SDG-2 rests on creating food systems that are more efficient, sustainable, and resilient, as well as more inclusive – of women, youth, small-scale food producers, and people living in poverty (HLPF, 2017). It is for such reasons that the 2030 Agenda emphasises the need for effective partnerships for the realisation of the sustainable development goals.

The monitoring and reporting of SDG-2 is an important process to ensure that the policies and laws put in place by countries yield the necessary impact as prescribed. According to HLPF (2017), monitoring and reporting are high on the agenda for many member states. The follow up and review process should be seen as an opportunity for governments and partners to inform and shape policy processes, share evidence on best practices and lessons learned, and concentrate efforts on proven solutions that deserve support. According to FAO et al. (2019), the proportion of undernourished people worldwide increased slightly from 10.6% in 2015 to 10.8% in 2018. This translates to 821 million people worldwide in 2018, up from 785 million in 2015 (FAO et al., 2019).

In 2018, 151 million children under age 5 suffered from stunting (low height for their age), 51 million suffered from wasting (low weight for height), and 38 million were overweight (Development Initiatives, 2018). In addition, stunting among children under 5 years of age is declining, albeit slowly, with the overall prevalence down from 32.6% in 2000 to 22.2% in 2017. Regionally, Asia has declined from 38.1% to 23.2% since 2000 and Latin America has declined from 16.9% to 9.6% (Development Initiatives, 2018). The prevalence of stunting has also declined in Africa from 38.32% to 30.3% since 2000 (Development Initiatives, 2018).

Of the 150.8 million stunted, 58.7 million live in Southern Asia. It is the same with wasting – more than half the children with wasting, 26.9 million, live in Southern Asia. Of the 38.3 million overweight children, Southern and East Asia have 5.4 million and 4.8 million children grappling with early overweight onset, respectively (Development Initiatives, 2018). While adult overweight and obesity is not tracked within the SDG framework, it is central to the nutrition agenda. Two billion adults are overweight, of whom 678 million are obese (NCD Risc, 2019). Globally, women have shown a higher prevalence of both overweight and obesity compared to men every year since 1990. Data on the prevalence of overweight adults increased from 35.7% in 2010 to 38.9% in 2016. Obesity prevalence in adults increased from 11.2% in 2010 to 13.1% in 2016 (Development Initiatives, 2018).

There is a global consensus that multi-dimensional, disciplinary, and sectorial approaches are essential to address malnutrition. These include a minimum intervention in the agricultural, health, educational, and environmental sectors that allows for an effective functioning of nutrition in food, health, water, and sanitation systems. Multi-dimensional approaches will also be needed to address the complexity of the multiple forms of malnutrition. The after-effects of malnutrition have different impacts and consequences depending on the specific population. Poor food security is causing millions of children to be stunted or too short for their age due to severe malnutrition. In this case, it is very important to rethink how food is grown, shared, and consumed. If done correctly, agriculture, forestry, and fisheries can provide nutritious food for all and generate decent incomes while supporting people-centred rural development and protecting the environment.

The strength of SDG-2 is that it combines agricultural productivity, conservation, and sustainable use of biodiversity and sustainable food systems to effectively address the objectives of the goal. Here, the focus is broadened to a more holistic approach to food systems integration that incorporates not only agriculture, diets, and nutrition, but the many elements across the food system, for example, storage, marketing, transport, processing, etc., and the drivers that impact and shift food systems. However, such integration requires a more complex multi-dimensional approach with solutions not only addressing hunger and nutrition but also going well beyond food availability and access to food through poverty reduction and social protection.

One of the major challenges of SDG-2 is that it requires complex, interlinked solutions across different levels, for example, urbanisation, rural transformation, water and forest conservation, infrastructure and technology development, sustainable agriculture, investment, etc. In addition, the interlinkages between SDG-2 and other SDGs are critical, mainly because of the interdependency of their targets, which is difficult to break. For example, the targets of SDG-1 (eradication of poverty) and SDG-2 are directly related to health, sustainable energy, biodiversity, sustainable cities, water, mitigation, adaption to climate change, etc. Furthermore, the challenges that SDG-2 faces are sometimes a stigma in communities. For example, the use of manure (e.g. animal remains) for crop cultivation is not viewed as sustainable and is prohibited. The agricultural sector is a broad sector that links politics, leadership, culture, and a healthy environment. Achieving SDG-2 will require coordination with international norms while having a mutual understanding of the farming system in a given country.

The progress on mobilising the means of implementing SDG-2 and the related provisions in the Addis Ababa Action Agenda, especially investment in sustainable and inclusive agriculture to reduce poverty and food insecurity, remains insufficient and needs to be urgently addressed (Daurte, 2017). Climate change continues to put pressure on existing resources. In the Horn of Africa, the main challenges have been climate change-induced impacts such as water shortages, droughts, desertification, soil erosion, and flooding (Shingirai & Tirivangasi, 2017).

There is a rapid degradation of soils, freshwater, oceans, forests, and biodiversity. The risks associated with disasters, such as droughts and floods, have increased in recent times and, as a consequence, have contributed to the loss of livelihood for many rural women and men, thus, forcing them to migrate to the city in search of more opportunities in a greener pasture. The use of more effective, inclusive platforms (including gender-inclusive and multi-stakeholder platforms and partnerships) are needed to support a holistic and well-balanced approach to achieving SDG-2 and the related targets.

In Africa, most countries are affected by civil war, drought and desertification, or lack of political leadership. For example, in Zimbabwe, hunger has been rampant because of the lack of strategy on how to sustainably manage an economy crippled by corruption and political instability. The situation became worse when the government refused to recognise land tenure rights, leaving most commercial farmers with no guaranteed land rights. As a result, it has become difficult for farmers to grow crops on their own land even though at times, in some regions, there is the investment and technology to sustain an advanced agricultural sector.

In some parts of the world, as is seen in Latin America, food insecurity can be caused by an influx of people, such as refugees or migrants. However, due to inequality and a vulnerable middle class, food insecurity in developing countries is always a problem. The most challenging part of achieving SDG-2 for many developing countries is that it demands many resources, significant effort, financial resources, and expertise, which are usually not available. This leaves many NGOs and financial institutions with the duty to reduce hunger. Generally, there is a very large population requiring food and financial assistance. The focus now is a situation where food aid and donors have to deal with an increased number to feed. Furthermore, the lack of development in many parts of world has left children and the most vulnerable with food that has little nutritional benefit for their health. The use of poor farming methods has also been identified as a challenge in achieving SDG2. It requires a collective effort to maintain a sustainable agriculture sector, and, in many countries, these efforts have been compounded by a slow global economy and rapid population growth. For example, major efforts meant to address sustainable agriculture will be hampered by climate change-induced impacts, which requires a collective effort to mitigate. This will lead to a decrease in agriculture productivity or limits in the availability of certain crops usually required by certain communities to reduce malnutrition.

In the context of rapid urbanisation, accelerating efforts around SDG-1 and SDG-2 requires efforts to improve urban–rural connections through investment in food systems. An improvement in interlinked food systems has the capability to increase urban market supplies with nutritious and affordable food, thus, expanding the economic opportunity for rural and peri-urban populations along the supply chain. Poor countries have engaged less in the discussions on developing indicators; agreements on indicators were not being followed by measuring the data provided. There are also gaps in the data collected in many developing countries to ensure that more solutions and funds are provided to achieve SDG-2.

As already mentioned, not all targets have the same degree of priority in different countries, which makes the concept of sustainability aligned with the SDGs very debatable. SDG targets and indicators should always be adjusted over time to accommodate socio-economic and institutional changes. The current limitation in data availability necessitates the need for a more holistic policy approach in the implementation of the SDG agenda. The challenges presented above have also opened up a wide range of research avenues, which could be beneficial in reducing the existing complexity of the SDG framework.

Governments across the world should take ownership of their developmental agenda and design national frameworks, policies, and measures to implement the SDGs in their respective countries. However, creating a more sustainable world as prescribed by the SDGs would require individuals to become sustainable change-makers. To this effect, specialised education focussed on sustainable development is required to empower change-makers and learners to make informed decisions and take responsible action for present and future generations (UNESCO, 2017). Therefore, education in the required knowledge, skills, values, and attitudes is a critical transformation to get on the path of sustainable development and to achieve the SDGs.

6

EXAMPLES – BEST PRACTICES IN ACHIEVING SDG-2

There is growing consensus around sharing ideas, knowledge, and skills to address common challenges and build shared strengths. Developing countries are increasingly recognising good practices from other developing nations as possible alternatives to accelerate progress in the achievement of the sustainable development goals (SDGs). 'The new surge in South–South and triangular cooperation has also seen countries in the South increasingly championing solutions that address common transnational development challenges which would be difficult to tackle independently' (United Nations Office for South–South Cooperation (UNOSSC), 2018). This chapter presents a series of good practices that can accelerate progress towards the implementation and achievement of the SDGs in developing countries. This series features seven examples of good practices collected from countries around the world. The presented examples are summarised from UNOSSC (2018), Parra, Dinsmore, Fassina, and Keizer (2015) and the United Nations Department of Economic and Social Affairs (UNDESA) (https://sustainabledevelopment.

un.org/partnerships/goodpractices). 'This compendium of good practices presents notable solutions at the national, sub-regional, regional and global levels to crucial challenges faced by developing countries ranging from efforts to eradicate poverty, reduce inequality, support climate change action and create peaceful and cohesive societies' (UNOSSC, 2018).

Five years into the implementation of the 2030 Agenda and the SDGs, many governments, UN entities, international and regional organisations, major groups, and other stakeholders have gathered sufficient evidence of progress in the implementation of the SDGs. Some of the critical questions raised by the United Nations Department of Economic and Social Affairs (UNDESA) in the process of gathering case studies include: 'What are the inspiring breakthroughs and success stories that are showing results and impacts?'; 'What are the good practices that can be replicated and scaled up?'; 'What are the gaps and constraints and how should we address them?'; 'Looking ahead, what steps should we take to accelerate progress?'

If the world is going to achieve the SDGs by 2030, countries need to be inspired by the examples of others, share good and best practice, and think of innovative ways to achieve the common goal of a sustainable future (International Centre for Trade & Sustainable Development (ICTSD), 2018). The evidence gathered in this chapter aims at benefiting developing countries seeking to scale up the implementation of SDG-2, particularly in the widespread application of policies, strategies, and programmes. The examples are concise, engaging, readable, and equally appealing to practitioner and non-practitioner audiences. Each case study has a brief background of the situation, strategic approach, results, and impact of the initiative, and a particular focus on challenges, lessons learned, and the potential for sustainability.

Example 1: Mozambique – Irrigated and Integrated Agro-production Systems Help Mozambique Adapt to Climate Change

The unpredicted climatic situation in Mozambique frequently causes extreme weather events such as drought, flood, and cyclones affecting the agricultural livelihood of the rural and the costal population. The country suffered 6 droughts and 15 floods between 2000 and 2009 alone. About 3.2 million people were affected by the droughts and 6 million were affected by floods. In 2010, both droughts and floods hit the country once again. The double shock left 465,000 people in need of food assistance, and 30% of cultivated land wiped out. Food insecurity became the norm in a country that already had a 37% undernourishment rate.

Despite the availability of water resources such as the Limpopo River, irrigation was limited, with most agricultural land around the Limpopo River depending on rainfall, thus, increasing the vulnerable of farms to unpredicted water supply. The region witnessed food insecurity, and this forced farmers to turn to the forest for food and income. Forest resources were consequently over-exploited and dwindled.

The *Environment mainstreaming and adaptation to climate change joint programme* under the framework of the MDG achievement fund focussed on addressing the issue of unpredictable rainfall in the remote district of Chicualacuala on the Limpopo River. The joint programme established by the UN leveraged on existing water resources to help the farmers in this semi-arid region to cope with climate change. The programme worked through the agricultural department of the Chicualacuala district government to promote an expansion of irrigated production in areas close to the Limpopo River. Farmers organised themselves around associations to better facilitate assistance from the joint programme. The programme staff collaborated with extension staff and community members to implement production activities. A community by community-based approach was used for the implementation of the programme. At each site, the programme staff, officials from the local government, and community members discussed ways to improve food security and climate change adaptation. Irrigated production was identified as a priority given the proximity to the river.

At the community level, the beneficiaries planned and executed activities including water harvesting, agriculture, irrigation, livestock production, integrated fish farming, pig keeping, forestry management, and the creation of natural resource management committees. Women were fully involved in these activities; in some activities such as agriculture, fish farming, and pig keeping, women formed the majority of the workforce. In each community, interested families signed up to join the farmer associations. The programme provided barbed wire to protect cultivated land from the intrusion of farm animals. Community members provided the labour.

The programme provided the initial supply of seeds and agricultural chemicals. Irrigation pumps and tubes were also supplied free of charge. The agricultural production practised and maintained by about 200 families has increased over time. This increased income, improved diet and, importantly, reduced pressure on forest resources. The use of manure, mulching, and small irrigation schemes became traditional practices among many farmers.

With new sources of income, farmers were able to improve the roofing of their homes, send their children to boarding schools, and have a sense of community and social gathering. Diversification of livelihood measures by using irrigated systems that combined crops, livestock, and forest resources helped the Chicualacuala district adapt to climate change.

The programme contended with internal disputes between certain influential members of the participating associations.

These disputes often had a negative impact on the production and the management of the fields. The programme overcame this problem by providing collaboration training to farmers' associations. The programme encountered challenges marketing the crops, particularly vegetables, which were mainly cultivated during the winter period. Over 400 tonnes of vegetable crops were produced in Chicualacuala between 2009 and 2011. However, because of overproduction in 2009 and coupled with poor market access, tonnes of vegetables were ruined.

To prevent the future spoilage of harvested vegetables, the programme engaged in training farmers in agro-processing techniques such as drying and a method to preserve excess harvest. Furthermore, the programme provided tractors, which increased the cultivated area, helping facilitate the marketing of products. A far-reaching innovation of the programme was the introduction of integrated fish farming within the fenced crop growing areas. The fish ponds were stocked with tilapia, and, on the banks of the fish ponds, pig sties, duck pens, and rabbit cages were built. This created an integrated production system based around crop growing activities, from which waste products and spoiled produce fed fish and other animals.

Manure from the animals and the fertilised water from the fishponds were returned to the fields. The integrated fish farming system created further challenges since the access to market was not considered in the overall planning. This oversight in planning could be attributed to the fact that fish farming was viewed by the programme staff and government officials as a subsistence activity and not a commercial activity. Irrigation tanks needed restocking every other week due to evapotranspiration and seepage. Given that using tanks was a fairly new practice in the region, the programme should have made it a priority to communicate the benefits to the farmers so as to ensure continued use, particularly in the case of using pumping systems instead of gravitational systems to stock the ponds. The overall management of the farming system became complex, and this resulted in some communities eliminating fish production and concentrating solely on agro-forestry and other livestock activities.

Example 2: Mexico – Supermarkets, Government Policy, and Changing Diets

According to Biles (2006), supermarkets in Mexico have been a fixture of the urban food retail economy since the late 1940s. The introduction of liberalisation policies in the 1980s that allowed foreign supermarkets to enter the market saw a growth in supermarkets in Mexican cities (Biles, 2006). Economic liberalisation intensified with the North American Free Trade Agreement (NAFTA) of 1994. According to Clark et al. (2012), NAFTA changed the types of products that were imported into the country by introducing high fructose corn syrup-based sweeteners that replaced cane sugar sweeteners, processed foods, and junk foods, as well as more dairy products, refined sugars, and ready-made meals. This saw a profound reshaping of the Mexican diet (Clark et al., 2012). Furthermore, the price for staple foods, such as maize (the basic ingredient in tortillas), has increased compared to fattier alternatives, resulting in a widespread move away from healthier traditional foods (Wiggins et al., 2015). Informal markets in the cities are selling primarily local and unprocessed foods, which help to support the traditional diet and provide nutritious food options (Rajagopal, 2010).

The food security issues in Mexico has been more about health and nutrition rather than the availability of food. This appears to be a food sovereignty issue, whereby the food system in Mexico is now based on imported, highly processed food; this is causing health issues (González, 2010). Over the past two decades, a notable trend of increasing type 2 diabetes rates, as well as Mexico's newfound status as the most obese country in the world, has caused concern (Clark et al., 2012). 'There was a 12 per cent increase in the obesity rate between 2000 and 2006, and by 2012, 39.7 per cent of Mexican adults were overweight and 29.9 per cent obese' (Clark et al., 2012). Furthermore, the first 5 years of NAFTA saw an increase in type 2 diabetes from 8.8% to 11.4% (Jiménez-Cruz & Bacardi-Gascon, 2004). Poor nutrition and the resulting health issues are the major concerns for food security in Mexico.

The Mexican government has started to put in place initiatives to combat poor nutrition and health issues. One of these initiatives is to improve children's health by using educational campaigns and implementing laws in regard to school children's diet (Castellanos, 2014). In 2014, a sales tax was introduced on soft drinks and beverages sweetened with high-fructose corn syrup. The additional revenue from the tax was then used in social programmes combating poor nutrition and the poor health situation in Mexico (The Economist, 2014). Indirectly, the tax imposed on poor-quality food affected the poor negatively, since such food is mostly purchased by the poor. The government, however, maintains that action is justified by the diet-related health problems in all socioeconomic groups. According to Parra et al. (2015), it is too early to determine if government policy has contributed to curbing demand for unhealthy foods, which would help show progress in the achievement of SDG-2.

Example 3: China – Wet Markets, Supermarkets, and Government Policy

'Chinese urban consumers traditionally have a cultural preference for fresh vegetables and fruits acquired through wet markets operated by independent informal vendors, leading to relatively healthy diets' (Li & Qian, 2010). According to Bai, Wahl, and McCluskey (2006), 60–70% of urban consumers were found purchasing most of their fruits and vegetables from wet markets, and only 10–15% from supermarkets. The current trend shows a rapid growth in overweight and obese citizens in Chinese cities (Parra et al., 2015). The increasing trend has been attributed to two main factors which include (a) the increase in access to processed foods (Hu, Reardon, Rozelle, Timmer, & Wang, 2004), and (b) the decrease in the prevalence of wet markets due to economic development policies. Generally, the price for fruits and vegetables is relative cheaper in wet markets and other informal food vendors than in supermarkets, thus, eliminating the chances for competition (Wang & Shi, 2012). The national government, however, hopes that the private sector will get more involved in food supply, a sector previously run by the public sector through wet markets (Parra et al., 2015). A certain number of wet markets are maintained across cities in China through the implementation of a zoning plan that sees various public facilities, such as hospitals, police stations, and wet markets constructed in designated locations (Zhang & Pan, 2013). Zoning regulations are often violated or ignored by the local population. In addition, wet markets tend to lower the value of a construction project because they are viewed as places with a lot of consumer traffic that are noisy and dirty. Consequently, developers tend to prefer reserving wet market space for supermarkets instead (Zhang & Pan, 2013).

The reduction of wet market space in urban areas has led to an increase in fees for food vendors to operate their market stalls. As a result, food vendors are forced to either raise food prices or exit the markets and engage in street hawking. The hawking of foodstuffs in the street is perceived as a nuisance because the vendors block traffic, there is noise pollution, and contamination from food waste; as a result, street hawking has been either banned or regulated in Chinese cities (Zhang & Pan, 2013). There is a positive correlation between the spatial density of wet markets and children's nutritional intake (Zhang & Pan, 2013). The decrease in the number of wet markets has contributed to the reduction of access to healthy foods, thereby granting more access to supermarkets with unhealthy food. It is essential to ensure equitable access to nutritious food by all, in particular, the poor and people in vulnerable situations, including infants, as is stipulated in Target 2.1 of SDG-2.

Example 4: Nairobi, Kenya – The Informal Economy and Access to Healthy Food

The rapid urbanisation rate in Kenya has made it absolutely necessary to address food access in urban areas. The current distribution of supermarkets in Kenya has helped significantly to improve food access for middle- and high-income urban households (Ouma, Mwangi, & Oduk, 2013), who can afford the high prices charged for fresh fruits, vegetables, and staple starches, such as maize, in supermarkets (Tschirley, Muendo, & Weber, 2004). The informal food markets are key players in providing access to more affordable healthy food for the local population in the capital city of Nairobi. According to the Rockefeller Foundation (2013), about 80% of food budgets are spent in the informal economy. An estimated 2.2 million livelihoods, constituting the majority of the 3.1 million inhabitants in Nairobi, rely on this economy.

It is, therefore, foreseen that with the consolidation of supermarkets in Kenya, healthier foods will be made available to the local population. Unfortunately, this does not translate into more access to healthy food for the urban poor because they cannot afford the high prices in supermarkets (Olack et al., 2011). In such conditions, the informal food economy will help to maintain affordable food prices for the poor, thus, increasing their food security as prescribed by SDG-2.

Kenya ranks 80th out of 109 countries as one of the most food insecure countries in the world (The Economist, 2014). The national and local governments have taken a practical approach in addressing food insecurity in Kenya (The Economist, 2014). Through the Economic Recovery Strategy for Wealth and Employment Creation, the national government has broadened the tax base to include the informal economy (Ministry of Planning and National Development, 2003), thus, creating a favourable environment for business. On the other hand, and in line with government action, the municipal government in Nairobi has provided infrastructure and other services, such as an open-air market to house 8,000 informal food vendors in the city centre, in return for a small tax (David, Ulrich, Zelezeck, & Majoe, n.d.). The policy has been appreciated for promoting sustainable access to food through an informal economy, even though the objectives were not geared towards urban food security. It is also important that these government policies address non-food issues such as transportation, finance capital, spatial inequality, and political violence, since they can all contribute to preventing the urban poor accessing the informal economy promoted by the government. 'Excluding the urban poor as a critical group of stakeholders exacerbates vulnerability and allows hazards to override other improvements of food security' (Parra et al., 2015).

Example 5: Chennai, India – The Role of the Government in Food Access and Distribution

Indian cities are extremely diverse in terms of food consumption, food sources, and food insecurity. Chennai is a city in India, located in the southern province of Tamil Nadu.

The city has a population of about 4.7 million, accounting for 13.5% of Tamil Nadu's urban total (Government of India, 2011). It is the fourth largest metropolitan agglomeration in India and one of the fastest-growing metropolitan areas. The population grew by approximately 7.7% between 2001 and 2011 because of internal migration; this has led to 26% of the population being slum dwellers (Harriss-White, Olsen, Vera-Sanso, & Suresh, 2013). The city is faced with challenges such as unemployment, poor infrastructure, inadequate city planning, and abject poverty; these have serious implications for health, sanitation, and food security. Food insecurity is classified as 'alarming rates of hunger', which according to the International Food Policy Research Institute is the classification used to for levels of hunger above serious.

In the last few years, supermarkets have become more prominent in Chennai. This does not help mitigate food insecurity because the poor cannot afford to use supermarkets. The average income of shoppers at supermarkets in Chennai is estimated at 340,000 rupees (US$7,649), and this is considerably higher than the average income in the province, which ranges between 40,000 and 50,000 rupees (US$900–$1,125) (Kattumuri, 2011). Consequently, the vast majority of slum dwellers living below the poverty line in the city are forced to depend on the government's 'Public Distribution System' (PDS) as their primary source of food. Since 2011, the PDS has offered free rice, subsidised sugar, and wheat to the poor residents who qualify for the scheme. The PDS has been described as one of the largest policy instruments in the world protecting the calorie intake of the poor, thus, improving their access to food and better health (Rajagopalan, 2010). It is important to state that there are situations of extreme undernutrition in the area, and the prevalence of calorie undernourishment is over 25% (Kattumuri, 2011). According to The Gazette of India (2013), the PDS has contributed to improving food security. This has resulted in the expansion of PDS programmes with the 2013 National Food Security Act. It has been estimated that the new programme will provide food to more than two-thirds of the Indian population, thereby helping to curb the classification rate 'alarming rates of hunger' (Menon, Deolalikar, & Bhaskor, 2009). Supermarkets have a significant role to play in the provision of food in the city of Chennai; however, many of the people in the urban slums rely on the government for food since they cannot afford the prices in the supermarkets. Government support, therefore, goes a long way to meeting some of the goals of the SDG.

Example 6: Brazil – Land Use Management

Brazil is a country with a very diverse economy and agro-climate with an agricultural production system that varies in terms of scale, intensification, and diversification (Gil et al., 2019). As reported by the latest census, family-based agriculture is conducted by about 80% of rural households; however, this represents only 24% of the total agricultural area (IBGE, 2006). According to Chaddad (2015), the census results illustrate the contrast with the reality of the three major hubs of commercial agriculture, including the South-East (export-oriented crops; vertically integrated agribusiness), Centre-West/Mapitoba (grazing, grain and fibre production; commercial and corporate farms; large-scale, industrialised farms), and South (mostly small-holders; diversified agriculture; cooperatives and contract farming).

Currently, Brazil is the third largest exporter of agricultural goods and is likely to become the first by 2024 (Organisation for Economic Co-operation and Development (OECD), 2015). This is mainly because there is an increasing demand for food and feed, both domestically and abroad (Gil et al., 2019). The main products include soy, beef, coffee, sugar, oranges, ethanol, and poultry (Companhia Nacional de Abastecimento

(CONAB), 2014). According to MAPA (2016), official statistics from 2015/16 to 2025/26 predict a 30% and a 29% increase in grain and meat production, respectively. Climate change losses are expected to be relatively small (Assad et al., 2013). A greater concern is the impact of Brazilian agricultural practices on the global climate, which is foreseen to intensify livestock production and the need to invest in land use mitigation actions (La-Rovere et al., 2014) 'such as restoration of degraded lands and farming diversification' (Gil et al., 2019). It should be noted that the impact level of the SDG-2 targets is very much interlinked and requires evaluation to identify the theories of change that explain causal pathways to impact, assumptions, and multi-directional effects.

Despite the fact that food production has significantly improved in recent decades, Brazil still experiences a malnutrition burden among its under-five population. Improving water productivity and the use of nutrients efficiently helps the country perform better and reduces the yield gap with other crops. Better land management practices and smart nutrient application should be promoted in areas with high rainfall and low nutrients, respectively. Soil quality was seen as declining, especially in areas with severe land degradation, such as pasture lands abandoned due to deforestation. This has contributed to low earnings for several farmers, especially farmers in the small-holder sector, thus, preventing them achieving Target 2.3 of SDG-2.

Example 7: Zimbabwe – Land Reform, Political Issues, and Agricultural Production

Zimbabwe saw a high production in the agricultural sector between the early 1980s and 1990s, and was referred as the breadbasket of the Southern African Development Committee (SADC) (United Nations, 2015a). In the past, there was a regular supply of maize to all Zimbabweans and other countries in the SADC region. Greater challenges of food supply were faced in the late 1990s because of the land reform crisis and politics; this has been a long debate in the country. According to many in the African continent, land reform in Zimbabwe was wrongly implemented in some parts of the country. Some landowners refused to vacate the land; for others, there were continuing conflicts between those allocated land and those who claimed ownerships of the land, hence, reducing food and agriculture production. Zimbabwe would have maintained its capacity for food production only if citizens had fully understood the land reform programme and political issues. There was an improvement in agricultural production starting in the early 2000s but with the occurrence of climate change, production continues to be affected negatively. Small-holder farmers are unable to produce 1.0 t ha^{-1} of maize because of climate change, low rainfall, and poor farming methods.

According to Nyamadzawo, Wuta, Nyamangara, and Gumbo (2013), the introduction of better farming methods, such as the use of rainwater harvesting techniques, has significantly improved maize yields in Zimbabwe. Furthermore, the adoption of nutrient intensification and rainwater harvesting in regions IV and V of Zimbabwe has the potential to improve food production as well (Nyamadzawo et al., 2016).

The use of improved sustainable agricultural production methods may have positive consequences for the environment, for example, through the rehabilitation of degraded lands, increasing vegetation cover, improved soil fertility, and sustainable land management systems; all these outcomes favour sustainable land use. In addition, sustainable agricultural production methods have the potential for reducing soil erosion and water pollution, and improving the number of trees in forests through practices such as agroforestry. Additional benefits of agroforestry are the sequestration of carbon

(Leakey, 2017) and an increase in fresh air (oxygen) in the atmosphere through photosynthesis.

Example 8: Nigeria – Agricultural Production and Its Effects on Economic Growth

The economy of about 90% of African countries depends on agriculture (United Nations, 2015a). Most farmers in Nigeria practice subsistence agriculture and only surpluses are sold; they operate at low levels depending on the size of the household. According to Gil et al. (2018), only a marketable surplus of up to 25% was detected based on household size. About 90% of agricultural output in Nigeria is produced by small-scale farmers using unsustainable techniques such as monoculture, which causes soil degradation. Soil degradation affects crop yields negatively because of the lost soil nutrients through leaching (Zhang et al., 2015). Maize productivity is estimated at 2t ha^{-1} and this is very low for Nigeria, especially when compared with other countries within the same climatic zone. With the amount of rainfall in Nigeria, maize production is expected to be high. Furthermore, the yield gap for specific crops in the country is estimated to be as high as 80% of the potential yield, thus, showing poor agricultural production (Gil et al., 2019; GYGA, 2018; World Bank, 2014).

The agricultural sector in Nigeria is tied to macro-developmental issues needing significant attention. Low crop production in the country is caused by land fragmentation, high vulnerability to climate shocks, and poor agricultural services. Land fragmentations have contributed to a reduced mechanisation of farmland, high transaction costs, and poor agricultural services, such as limited infrastructure, technical assistance, access to credit, and access to fertilizers, which has affected crop yield negatively (Manyong, 2005). Food importation in Nigeria has increased by 11% per year on an average

(Gil et al., 2019) and this amounts to about US$600 billion per year (World Bank, 2014).

The population of Nigeria is projected to be the third largest in the world by 2050 (Van-Ittersum et al., 2016), thus, increasing the demand for food. Food security might be achieved through the use of fertilisers, appropriate technology, and mechanisation, thus, achieving zero hunger and reducing poverty in Nigeria. The increased use of irrigation schemes in small-holder farming systems and increasing farm sizes of small-holder farmer might also contribute to improving food security, access to nutrition, and reduce malnutrition in the under-fives. To improve crop yields and food security, agricultural production must be intensified in a sustainable manner. The yield gap must be reduced, thus, alleviating pressure on food imports and improving access to cheaper food resources, nutrition, and water (Gil et al., 2019). Finally, the empowerment of small-holder farmers could also be a better way of improving food access for the majority of the country and reducing food costs.

7

CONCLUSION AND RECOMMENDATIONS

7.1 CONCLUSIONS

The agricultural sector is key in achieving the United Nations Sustainable Development Goals. However, it is very important to have an understanding of the interactions with other sectorial actions so as to ensure coherence and maximise the benefits shared with other sectors. For example, local agricultural production and school feeding programmes could generate win–win situations through income earnings for small producers and their families, alongside nutritional and cognitive gains (and likely future income) for school-age children. Furthermore, existing social protection programmes could protect the nutrition and health of poor small-holder households during periods of high seasonal variation and climate shocks, leading to poor yields.

Malnutrition in all its forms (undernourishment, micronutrient deficiencies, and overweight) is a reality in many countries, especially in developing countries. Malnutrition has robbed communities of good health, increased productivity, improved human resource capacity, economic growth, etc.

Interventions to address malnutrition are often considered as falling under the health sector, thus, undermining the synergies and benefits that come with integrating such actions within the agricultural and other sectors for more coherence. Countries should continue to ensure that the agricultural and corresponding sectors become more nutrition sensitive so that they drive the much-desired nutrition revolution needed to achieve SDG-2. More nutrition sensitive practices or programmes could be achieved by reviewing existing agriculture, food, and trade policies to identify necessary reforms to stimulate the local supply and demand of healthy nutritious foods, and discourage the consumption of unhealthy foods and food waste. In addition, practices or programmes should ensure that unfavourable food policies do not aggravate nutritional challenges, especially in rapidly transforming food systems. They should rather create and strengthen institutional and policy environments that enable agriculture to support nutrition and health goals. Developing countries should harness the potential for science, technology, and innovation to reduce post-harvest losses and food waste; promote product diversification with nutritious foods; improve processing to extend shelf life and make healthy foods easier to prepare; and improve storage and preservation to retain nutritional value, ensure food safety, and extend seasonal availability.

The promotion of sustainable food consumption patterns that ensure the availability of healthier and more nutritious foods at affordable prices for all, especially for vulnerable communities, is an important milestone for achieving SDG-2. A political choice should be made to position nutrition as a priority at the highest level of governance within an integral element of funded comprehensive growth and development strategies. Furthermore, future programme design should integrate nutrient objectives and indicators in their monitoring mechanism, seeking to achieve national priority developmental objectives. These new initiatives will require the establishment of strong institutional structures to coordinate the recommended efforts and also to ensure that existing resources in agriculture, social protection, education, and water and sanitation are leveraged to scale up nutrition impact. Furthermore, there is a need to promote national growth and development strategies including a blend of nutrition-specific and nutrition-sensitive programmes that seek to increase the overall supply and distribution of healthy nutrient-dense foods at affordable prices through agricultural value chains that support sustainable livelihoods for rural households.

There is also a need to develop capacity and leadership to use evidence-informed decision making to enhance the impact of agriculture on nutrition and the health of the local population. Leadership is a sine qua non in effectively coordinating and implementing the recommended nutrition programmes and interventions. In addition, the long-term sustainability of the programmes and interventions will be determined by the available workforce trained by local institutions. This will also help to leverage current momentum on nutrition sensitivity into the future. Good evidence is built on the availability of a good database. It is crucial to invest in data collection, which will be used to support annual national reporting on progress made in the implementation of SDG-2.

7.2 RECOMMENDATIONS

7.2.1 Knowledge, Communication, and Evidence

One of the most important enabling environments needed to achieve SDG-2 is the ability to harness knowledge, promote communication, and provide evidence of the progress made on issues affecting implementation at the national level and worldwide. Often, there is a serious disconnection between SDG-2 responses and targets to be achieved, thus, contributing to inaccurate reporting of the achievement. The collection of integrated and national data across agriculture, food systems, nutrition intake, health, etc., in line with input data for the respective indicators and targets, is highly recommended. Only with sufficient and good data are countries able to transparently report their progress in regard to the implementation of SDG-2, as well as to make informed decisions on their strategy to end hunger and food insecurity.

7.2.2 Politics, Governance, and Policy

Policies should have clear objectives that derive from a comprehensive analysis of the gaps and the overall weaknesses that surround food systems and the SDG-2 targets; note that the focus here is on policies addressing SDG-2 and the interlinkages to other SDGs so as to build synergies, ensure coherence, and sustainability. Within SDG-2, it is also important to coordinate agricultural production, market and trade systems, food transformation and consumer demand, consumer purchasing power, and other processes so as to drive the necessary change. The government of every country has the responsibility of providing the institutional and policy environments that cater for monitoring and ensuring progress. Confidence and sustainability will be achieved only when there is transparency and accountability at all levels of the system.

Governments should promote policies that support vulnerable people with access to existing systems and to benefit from the resources of the country. Furthermore, meaningful and long-term progress is achieved when the local population drives societal development. The government should promote domestic policies and investment that help transform food systems to reduce hunger, malnutrition, and improve the health of under-fives and the majority of citizens. The government should also invest in agricultural research, development, and innovation systems that target the overall objective of the SDGs, and, in particular, small-holder farmers. They produce about 90% of food in developing countries but are resource poor and unable to develop themselves. Through such investments, agricultural productivity will be improved, and small-holder farmers will contribute to reducing poverty and improving nutrition and food security, thus, reducing malnutrition and deaths from nutritional deficiencies at large.

Government should also design special policies that target and improve equity through participation, access to resources, and decent employment for vulnerable groups. Any action should be contextualised within a broader political perspective, for example, women's access to land, employment, health service, education, etc. Similarly, a gender lens should be applied to assess the participation of women in the agricultural sector and also, perhaps, assess the role and impact of women in mediating the nutrient level in agriculture and society. Generally, female empowerment should be strengthened in all spheres of the community. In addition, programmes addressing undernutrition should not put too much emphasis on quantity of calories and weight gain, while anti-obesity campaigns must avoid the unintended consequences of undernutrition. It is also important to incorporate explicit nutrition objectives and indicators into the design of agricultural programmes, and track and mitigate potential harms.

7.2.3 Agricultural Production

Agricultural production requires starting inputs such as machines, equipment, loans, knowledge, and support from agricultural extension services, etc. In most countries, the people often have resources such as land and are willing to engage in agricultural activities; however, they lack the capital to invest in agriculture, in the diversification of their activities, and accessing the market. It is, therefore, important to align agricultural research investments to the improvement of agricultural products (such as fruits, vegetables, animal source foods, nuts, seeds), facilitate access to market and better input/cultivation methods. This will promote diverse production systems and the diversification of income sources for small-holder farmers, thus, supporting the implementation of SDG-2. In addition, vulnerable groups such as women should be protected, be empowered to have access to loans, own land, and be provided with agricultural support. In most developing countries, women tend to spend more time on farms because of men migrating to towns in search of jobs or being more easily employed in other sectors. Environmentally friendly and sustainable production integrating other sectors should be promoted at all times.

7.2.4 Market and Trade Systems

Market and trade systems are not easily accessible by the local population, especially for small-scale farmers in developing countries. There is also a lack of infrastructure such as communication networks and roads to connect rural areas to local market in towns, as well as the means to connect to international markets. The government should support agricultural producers by developing infrastructures such as farmer to market roads, as well as facilitating their access to both local and international markets. However, the government, with their policies, should ensure a balance in the supply of diverse fresh foods between local markets and the international value market chains, thus, promoting access to food for all groups.

There is also the need to improve on food processing, storage, and preservation so as to retain nutritional value and food safety, reduce seasonality and post-harvest losses, and to ensure the transportation of healthy foods. Food price volatility is a major challenge for small-holder farmers in developing countries since they play little or no role in the determination of the prices of the foodstuffs they produce. There is, therefore, the need for better management of food price volatility, especially to protect vulnerable groups in having economic access, as well as ensuring their access to food resources from the markets.

7.2.5 Promotion of the Development and Use of a National Food-Based Dietary Guideline

The development of guidelines to use natural and national food-based diets will ensure healthy eating habits at both local and national levels. Governments should restrict the advertisement, marketing and commercial promotion of unhealthy, low nutrient, ultra-processed foods. This has the great advantage of promoting human and environmental health. Furthermore, putting in place mechanisms to control food labelling is a good approach to ensure that claims are evidence-based and the products' quality are maintained at all times. It is also necessary to prioritise the improvement of diet quality (e.g. to include animal source foods such as fish, meat, eggs, and dairy) in young children, adolescent girls, and women.

Schools and hospitals need to improve the quality of institutional diets (promoting the consumption of fruits and vegetables, legumes/pulses, nuts and seeds, high-protein, micronutrient-dense grains, and safe milk) and to ensure a balanced diet, especially for primary schools, since they are still growing and are exposed to threats of disease resulting from nutritional deficiency.

The designed guideline should consider the societal pathways between agriculture (as a livelihood) and nutrition-relevant outcomes (not only child stunting), especially pathways in which women are significantly engaged.

7.2.6 Capacity, Leadership, and Financing

Capacity at individual, organisational, and systemic levels and leadership are important frameworks to highlight the enabling environment for implementing SDG-2. Financial resources also play an important role in the implementation process, as well as strengthening the development of capacity with diverse disciplines at different administrative levels. There is also a need to cultivate leadership, promote nutrition champions, policy entrepreneurs, and civil society activists at all levels. In such a multi-disciplinary context, cross-sectoral leadership could be beneficiary in bridging sectoral divides (e.g. between agriculture and health, or between SDG 1 and SDG-2, as well as other SDGs), in budget allocations, and in setting the overall SDG-2 strategy at the national level.

In general, developing policies towards the achievement of SDG-2 should be driven by the experience of the local context. Particular attention should be given to the role of the informal economy, non-food issues, pro-poor pricing structures, healthy food consumption patterns (especially, in the urban context), and small-holder farmers in ensuring food security and the availability of food for all, including vulnerable groups such as women and children. The examples from Mexico, China, Kenya, and India have highlighted the food security challenges already faced by urban dwellers and how they are using existing food systems such as supermarkets, the informal economy, and PDSs to combat these challenges. Small-holder farmers have a very important role to play in the achievement of SDG-2, especially for developing countries. There are about 500 million small farms worldwide, most still rain-fed, providing up to 80% of the food consumed in a large part of the developing world (Parra et al., 2015). An important way to increase food security and nutrition for the poorest, as well as food production for local and global markets, is to invest in small-holder women and men farmers. Even though the local context is very important in driving the implementation of SDG-2, South–South cooperation is highly recommended for developing countries as a platform for a best practice approach from their peers to promote sustainable food security.

BIBLIOGRAPHY

Alkire, S., & Samman, E. (2014). *Mobilising the household data required to progress toward the SDGs*. Oxford: Queen Elizabeth House, University of Oxford.

Assad, E., Pinto, H. S., Nassar, A., Harfuch, L., Freitas, S., Farinelli, B., Fernandes, E. (2013). *Impacts of climate change on Brazilian agriculture*. Washington, DC: World Bank LAC & PROFOR.

Bai, J., Wahl, T., & McCluskey, J. (2006). *Consumer choice of retail food store formats in Qingdao*. China. Long Beach.

Ballard, T., Kepple, A., & Cafiero, C. (2013). *The food insecurity experience scale: Developing a global standard for monitoring hunger worldwide*. Technical Paper. Rome: FAO.

Barbier, E. B. (2016). Sustainability and development. *Annual Review of Resource Economics*, *8*, 261–280. Retrieved from http://www.annualreviews.org/doi/abs/10.1146/annurev-resource-100815-095227

Barbier E. B., & Burgess, J. C. (2017). The Sustainable Development Goals and the systems approach to sustainability. *Economics: The Open Access Journal*, *11*(2017–28), 1–23. Retrieved from http://dx.doi.org/10.5018/economics-ejournal.ja.2017-28

Biles, J. (2006). Globalization of food retailing and the consequences of Wal-Martization in Mexico. *Academia*, *24*, 343–355.

Cabell, J., & Oelofse, M. (2012). An indicator framework for assessing agroecosystem resilience. *Ecology and Society*, *17*, 18.

Carlson, K. M., Gerber, J. S., Mueller, N. D., Herrero, M., Macdonald, G. K., Brauman, K. A., ... Saatchi, S. (2016). Greenhouse gas emissions intensity of global croplands. *Nature Climate Change*, *7*(1), 63.

Castellanos, J. (2014). *2014 Mexico food processing report*. In Global Agricultural Information Network Report, 1–19. Retrieved from http://gain.fas.usda.gov/RecentGAINPublications/FoodProcessingIngredients_MexicoCity ATO_Mexico_12-9-2014.pdf

Chaddad, F. (2015). *The economics and organization of Brazilian agriculture: Recent evolution and productivity gains*. Cambridge: Academic Press.

Clark, S., Hawkes, C., Murphy, S., Hansen-Kuhn, K. A., & Wallinga, D. (2012). Exporting obesity: US farm and trade policy and the transformation of the Mexican consumer food environment. *International Journal of Occupational and Environmental Health*, *18*(1), 53–65.

Companhia Nacional de Abastecimento (CONAB). (2014). Se'ries Histo'ricas Relativas a's Safras 1976/77 a2013/14 de A' rea Plantada, Produtividade e Producao. Brasilia: Companhia Nacional de Abastecimento.

David, S., Ulrich, O., Zelezeck, S., & Majoe, N. (n.d.) *Managing informality: Local government practices and approaches towards the informal economy*. Pretoria: South African Local Economic Development Network, South African Local Government Association & Local Economic Development Network of Africa.

Daurte, E. C. (2017). Agriculture development, food security and nutrition. A statement on behalf of the group of 77 and China by the first secretary permanent mission of Ecuador to the UNITED NATIONS, at the second committee of the seventy second session on the UN General assembly on Agenda Item, New York, October 16. Retrieved from http://papersmart.unmeetings.org/media2/16153739/ecuador-77-e.pdf

Development Initiatives. (2018). *Global Nutrition Report: Shining a light to spur action on nutrition. Bristol, UK: Development Initiatives.*

EU. (2017). Agriculture and rural development 2014–2020—The Netherlands fact sheet. European Commission. Retrieved from https://ec.europa.eu/agriculture/rural-development-2014-2020/country-files/nl_en

FAO. (2008). An introduction to the basic concepts of food security. Food security information for action. Practical guides. EC—FAO Food Security Programme.

FAO. (2014). *Area equipped for irrigation. Infographic. AQUASTAT: FAO's information system on water and agriculture.* Rome: Food and Agriculture Organization of the United Nations (FAO). Retrieved from http://www.fao.org/nr/water/aquastat/infographics/Irrigation_eng.pdf

FAO. (2016a). *Food and agriculture: Key to achieving the 2030 agenda for sustainable development.* Rome: Food and Agriculture Organization.

FAO. (2016b). *Minimum dietary diversity for women: A guide for measurement.* Rome: FAO & FHI 360.

FAO. (2016c). FAOSTAT Database. Retrieved from http://faostat3.fao.org/download/Q/QV/E

FAO. (2016d). *The state of food and agriculture: Climate change, agriculture and food security, 2016*, Rome: Food and Agriculture Organization of the United Nations.

FAO. (2020a). Ghana at a glance. Retrieved from http://www.fao.org/ghana/fao-in-ghana/ghana-at-a-glance/en/

FAO. (2020b). Gambia at a glance. Retrieved from http://www.fao.org/gambia/gambia-at-a-glance/en/

FAO; ICRISAT; CIAT. (2018). Climate-Smart Agriculture in the Gambia. CSA Country Profiles for Africa Series. International Center for Tropical Agriculture (CIAT); International Crops Research Institute for the Semi-Arid Tropics (ICRISAT); Food and Agriculture Organization of the United Nations (FAO). Rome, Italy (20 p).

FAO, IFAD, UNICEF, WFP & WHO. (2019). *The State of Food Security and Nutrition in the World 2019. Safeguarding against economic slowdowns and downturns*. Rome: FAO. Licence: CC BY-NC-SA 3.0 IGO.

Federal Ministry of Agriculture and Rural Development (FMARD). (2012). *Agricultural transformation agenda*. Abuja: FMARD.

Gambia Bureau of Statistics (GBOS). (2016). Consumer Price Index (CPI), GBOS, The Gambia. Retrieved from www.gbos.gov.gm/uploads/naccounts/2012

Gambian Ministry of Finance and Economic Affairs. (2014). Level of Achievement of the MDGs: MDG Status Report 2014. Retrieved from www.gm.undp.org/content/gambia/en/home/library/poverty/mdg-2014-report.html

Gil, J. D. B., Reidsma, P., Giller, K., Todman, L., Whitemore, A., & van Ittersum M. (2019). Sustainable development goal 2: Improved targets and indicators for agriculture and food security. *Ambio*, *48*, 685–698. Retrieved from https://doi.org/10.1007/s13280-018-1101-4

Gillespie, S., & van den Bold, M. (2017). Agriculture, food systems, and nutrition: Meeting the challenge. *Global Challenges 2017*(1), 1600002.

Goldstein, B., Hauschild M. Z., Fernandez, J., & Birkved, M. (2016). Urban versus conventional agriculture,taxonomy of resource profiles: a review. *Agronomy for Sustainable Development, 36*(9. doi:10.1007/s13593-015-0348-4.

González Dávila, O. (2010). Food security and poverty in Mexico: The impact of higher global food prices. *Food Security*, *2*, 383–393.

GYGA. (2018). Global yield gap atlas. Wageningen University and Research, University of Nebraska—Water for Food. Retrieved from http://www.yieldgap.org/web/guest/home

High-Level Political Forum on Sustainable Development (HLPF). (2017). Thematic review of SDG2: End hunger, achieve food security and improved nutrition, and promote sustainable agriculture. High-level Political Forum on Sustainable Development. Retrieved from https://sustainabledevelopment.un.org/content/documents/14371SDG2_format.revised_FINAL_28_04.pdf

Harriss-White, B., Olsen, W., Vera-Sanso, P., & Suresh, V. (2013). Multiple shocks and slum household economies in South India. *Economy and Society*, *42*(3), 398–429.

Hák, T., Janoušková, S., & Moldan, B. 2016. Sustainable development goals: A need for relevant indicators. *Ecological Indicators*, *60*, 565–573.

Hu, D., Reardon, T., Rozelle, S., Timmer, P., & Wang, H. (2004). The emergence of supermarkets with Chinese characteristics: Challenges and opportunities for China's agricultural development. *Development Policy Review*, *22*(5), 557–586. doi:10.1111/j.1467-7679.2004.00265.x

ICTSD. (2018). *Achieving sustainable development goal 2: Which policies for trade and markets?* Geneva: International Centre for Trade and Sustainable Development (ICTSD).

IFPRI. (2015). Global Nutrition Report 2015: Actions and Accountability to Advance Nutrition and Sustainable Development. Washington, DC: International Food Policy Research Institute (IFPRI).

International Council for Science (ICSU). (2017). A guide to SDG interactions: From science to implementation. In D. J. Griggs, M. Nilsson, A. Stevance, D. McCollum (Eds.). Paris: International Council for Science. doi:10.24948/2017.01

International Food Policy Research Institute. (2015). *Global Nutrition Report 2015: Actions and accountability to advance nutrition and sustainable development*. Washington, DC. http://dx.doi.org/10.2499/9780896298835

International Labour Organisation. (2018). Transforming our world: A cooperative 2030 – Cooperative contributions to SDG 2. ILO Briefs, October 2018. Retrieved from http://www.socioeco.org/bdf_fiche-document-6644_it.html

Jiménez-Cruz, A., & Bacardi-Gascon, M. (2004). The fattening burden of type 2 diabetes on Mexicans: Projections from early growth to adulthood. *Diabetes Care*, *27*, 1213–1215.

Kattumuri, R. (2011). Food Security and the Targeted PDS in India." ARC working Paper 38 LSE.

Knox, J., Daccache, A., & Hess, T. (2013). What is the impact of infrastructural investments in roads, electricity and irrigation on agricultural productivity? Systematic Review CEE 11-007. London: International Initiative for Impact Evaluation.

La Rovere, E. L., Pereira, Jr., A. O., Dubeux, C. B. S., & Wills, W. (2014). Climate change mitigation actions in Brazil. *Climate and Development*, 6, 25–33.

Le Blanc, D. (2015). Towards integration at last? The sustainable development goals as a network of targets. *Sustainable Development, 23*, 176–187.

Leakey, R. R. B. (2017). *Multifunctional agriculture: achieving sustainable development in Africa*. San Diego, CA: Academic Press.

Li, T., & Qian, Y. (2010). Mainlanders visit HK for cheap groceries. *China Daily*, November 10.

Manyong, V. M. (2005). *Agriculture in Nigeria: Identifying opportunities for increased commercialization and investment*. Nigeria: IITA.

MAPA. (2016). Projecoes do Agronegocio—Brasil 2015/16 a 2025/26—Projec¸o˜es de Longo Prazo. 7 ed. Brasilia, Brazil: Secretaria de Politica Agricola—Ministerio da Agricultura, Pecuaria e Abastecimento—Government of Brazil.

Menon, P., Deolalikar A., & Bhaskor, A. (2009). India State Hunger Index, Comparison of Hunger Across States. *International Food Policy Research Institute*, 5–29.

Ministry of Planning and National Development. (2003). Kenya: Economic recovery strategy for wealth and employment creation 2003-2007. Government of Kenya.

Mollier, L., Seyler, F., Chotte, J. L., & Ringler, C. (2017). End hunger, achieve food security and improved nutrition and promote sustainable agriculture: SDG2: In: Stevance A. S. (coord) a guide to SDG interactions: from science to implementation (SDG: Sustainable Development Goals) (pp. 31–80). Paris: ICSU.

NCD Risc. (2019). http://ncdrisc.org/. Accessed on November 8, 2019.

Nepstad, D., McGrath, D., Stickler, C. Alencar, A., Azevedo, A., Swette, B., ... Digi no, M. (2014). Slowing Amazon deforestation through public policy and interventions in beef and soy supply chains. *Science*, *344*, 1118–1123.

Nilsson, M., Griggs, D., & Visbeck, M. (2016). Map the interactions between sustainable development goals. *Nature*, *534*, 320–322. Retrieved from https://www.nature.com/news/policy-map-the-interactions-betweensustainable-development-goals-1.20075

Nwachukwu, I. N. (2014). Dyanamics of agricultural exports in sub-sahara Africa: An empirical study of rubber and cocoa from Nigeria. *International Journal of Food and Agricultural Economics*, 2(3), 91–104. ISSN 2147-8988.

Nyamadzawo, G., Gotosa, J., Moyo, F., & Govere, I. (2016). The potential of tied contours and crop intensification and on maize yields in semi-arid smallholder farming area of Marange, Zimbabwe.

Nyamangara, J., Mudhara, M., & Giller, K. E. (2013). Effectiveness of cattle manure and nitrogen fertiliser application on the agronomic and economic performance of maize. *South African Journal on Plant and Soil, 22*(1), 59-63. doi: 10.1080/02571862. 2005.10634682

Nyamadzawo, G., Wuta, M., Nyamangara, J., & Gumbo, D. (2013). Opportunities for optimizing infield water harvesting to cope with changing climate in semi-arid smallholder farming areas of Zimbabwe. *Springer Plus*, *2*, 1–9.

Olack, B., Burke, H., Cosmas, L., Bamrah, S., Dooling K., Feikin, D. R., … Breiman, R. F. (2011). Nutritional status of under-five children living in an informal urban settlement in Nairobi, Kenya. *Journal of Health, Population and Nutrition*, *29*(4), 357–363.

Organisation for Economic Co-operation and Development. (2015). *Agricultural outlook 2015*. Paris: OECD/UN FAO.

Parra, D. C., Dinsmore, K., Fassina, N., & Keizer, C. (2015). Toward SDG 2: Food security and urbanization in the global south. Centre for International Governance Innovation Graduate Fellows Policy Brief no. 8. Retrieved from https://www.cigionline.org/series/cigi-graduate-fellows-policy-brief-series

Popkin, B. M. (2014). Nutrition, agriculture and the global food system in low and middle-income countries. *Food Policy*, *47*, 91–96.

Ouma, D., Mwangi D. T., & Oduk. M. P. (2013). Modeling agility in Kenyan supermarkets chain expansion. *International Journal of Business and Commerce*, 2(8), 21–36.

Rajagopal, V. (2010). Street markets influencing urban consumer behaviour in Mexico. *Latin American Business Review*, *11*, 77–110.

Rajagopalan, S. (2010). *The role of targeted PDS and food stamps in promoting better access to food in poor households*. Chennai: M. S. Swaminathan Research Foundation Report.

RIVM. (2015). *Greenhouse gas emissions in the Netherlands 1990–2013 National Inventory Report 2015*. The Hague.

Rosegrant, M., & Hazell, P. (2000). *Transforming the Rural Asian Economy*: The Unfinished Revolution. Hong Kong: Oxford University Press.

Rockefeller Foundation. (2013). The Rockefeller Foundation's informal city dialogues: Nairobi, Kenya. http://nextcity.org/informalcity/city/nairobi

Shingirai, S. M., & Tirivangasi, H. M. (2017). Climate change: A threat towards achieving 'Sustainable Development Goal number two' (end hunger, achieve food security and improved nutrition and promote sustainable agriculture) in South Africa. *Jamba*, *9*(1), 350.

Slätmo, E., Fischer, K., & Röös, E. (2016). The framing of sustainability in sustainability assessment frameworks for agriculture. *Sociologia Ruralis*, *57*(3), 378–395.

Steiner-Asiedu, M., Dittoh, S., Newton, S. K., & Akotia, C. (2017). Addressing Sustainable Development Goal 2: The Ghana Zero Hunger Strategic Review. The Full Report. Retrieved from file:///H:/UNU%20Work/SDG2/Ghana.pdf

Synthesis Report of the Nigeria Zero Hunger Strategic Review. 60 pp.

Szabo, S., Laperriere, A., Kripke, G., Bhalla, J., Kenei, S., Bahalim, A., & Bizikova, L. (2017). A collaborative process towards a SDG2 accountability framework. Global Open Data for Agriculture and Nutrition 1-17. doi: 10.13140/RG.2.2.18861.49126

Szabo, S., Mowlds, S., Claros, J. M., Kar, A., Knechtel, W., Di Ciommo, M., & Kashim, I. (2016). Towards a data users' framework to advance Sustainable Development Goal 2. *International Journal of Population Studies*, *2*(1), 1–12.

Szabo, S. Nicholls, R., Neumann, B., Renaud, F., Matthews, Z., Sebesvari, Z., ... Hutton, C. (2016). Making SDGs work for climate change hotspots. *Environment: Science and Policy for Sustainable Development*, *58*(6), 24–33.

Data for Agriculture and Nutrition, 1–17. doi:10.13140/RG.2.2.18861.49126

Tosun, J., & Leininger, J. (2017). Governing the interlinkages between the sustainable development goals: Approaches to attain policy integration. *Global Challenges*, *2017*, 1, 1700036. doi:10.1002/gch2.201700036

The Economist. (2014). Poverty in Mexico: A few more Pesos, a lot more stress. *The Economist*, January 4.

The Gazette of India. (2013). The National Food Security Act (2013). India. Retrieved from http://indiacode.nic.in/acts-in-pdf/202013.pdf

Tschirley, D., Muendo, K. M. & Weber, M. T. (2004). Improving Kenya's domestic horticultural production and marketing system: Current competitiveness, forces of change, and challenges for the future. Tegemeo Institute of Agricultural Policy and Development, Egerton University Working Paper 8B.

UNESCO. (2017). Education for Sustainable Development Goals: Learning objectives. UNESCO. Assistant Director-General for Education, 2010–2018 (Qian Tang). Writer of foreword [718]. ISBN: 978-92-3-100209-0, 978-92-3-600061-9.

UN General Assembly. (2015). Transforming our world: The 2030 Agenda for Sustainable Development, 21 October

2015, A/RES/70/1, Retrieved from https://www.refworld.org/docid/57b6e3e44.html

United Nations. (2015a). Global Sustainable Development Report, 2015 edition.

United Nations. (2015b). Adoption of the Paris Agreement. In *21st Conference of the Parties, United Nations Framework Convention on Climate Change* (UNFCCC), Paris.

United Nations. (2015c). *Transforming our world: The 2030 agenda for sustainable development*. United Nations, New York. Retrieved from https://sustainabledevelopment.un.org/post2015/transformingourworld/publication

United Nations. (2016). In *Global Sustainable Development Report* (2016 ed.). New York, NY: Department of Economic and Social Affairs.

United Nations. (2019). *Zero hunger: Why it matters*. New York, NY: United Nations. Retrieved from https://www.un.org/sustainabledevelopment/wp-content/uploads/2016/08/2.pdf

United Nations. (2020a). Goal 2: Zero Hunger. Sustainable development goals. Retrieved from https://www.un.org/sustainabledevelopment/hunger/

United Nations. (2020b). Global indicator framework for the Sustainable Development Goals and targets of the 2030 Agenda for Sustainable Development. Retrieved from https://unstats.un.org/sdgs/indicators/indicators-list/

United Nations Economic and Social Council. (2016). Progress towards the Sustainable Development Goals, Report of the Secretary-General, High-level political forum on sustainable development, convened under the auspices of the Economic and Social Council.

United Nations Economic and Social Council. (2019). Report of the Secretary-General. Special edition: Progress towards the Sustainable Development Goals E/2019/68.

United Nations Office for South–South Cooperation. (2018). *Good practices in South–South and Triangular Cooperation for sustainable development* (Vol. 2). New York, NY: United Nations Office for South–South Cooperation (UNOSSC). Retrieved from https://www.unsouthsouth.org/2018/09/12/good-practices-in-south-south-and-triangular-cooperation-for-sustainable-development-vol-2-2018/

United Nations Statistics Division (UNSD). (2016a). The SDGs report. Zero hunger. Goal 2. Retrieved from https://unstats.un.org/sdgs/report/2016/goal-02/

United Nations Statistics Division (UNSD). (2016b). *Zero hunger. UN Statistic Report*. New York, NY: United Nations. Retrieved from https://unstats.un.org/sdgs/report/2016/goal-02/

Valensisi, G., & Karingi, S. (2017). From global goals to regional strategies: Towards an African approach to SDGs [the Sustainable Development Goals]. *African Geographical Review*, *36*(1), 45–60. DOI: 10.1080/19376812.2016.1185738

von Stechow, C., Minx, J. C., Riahi, K., Jewell, J., McCollum, D. L., Callaghan, M. W., Bertram, C., Luderer, G., & Baiocchi, G. (2016). 2 oC and SDGs: united they stand, divided they fall? *Environmental Research Letters*, 11, 034022.

Wang, R., & Shi, L. (2012). Access to food outlets and children's nutritional intake in urban China: A difference-in-difference analysis. *Italian Journal of Pediatrics*, *38*(1), 30. doi:10.1186/1824-7288-38-30

Wiggins, S., Keats, S., Han, E., Shimokawa, S., Alberto, J., Hernandez, V., & Claro, R. M. (2015). The Rising Cost of a Healthy Diet: Changing Relative Prices of Foods in High-Income and Emerging Economies. Overseas Development Institute Report, May.

World Bank. (2017). Poverty and equity database. The World Bank, global poverty working group. Retrieved from http://povertydata.worldbank.org/poverty/home/

Zhang, Q. F., & Pan, Z. (2013). The transformation of urban vegetable retail in China: Wet markets, supermarkets and informal markets in Shanghai. *Journal of Contemporary Asia*, *43*(3), 497–518. doi:10.1080/00472336.2013.782224

Zhang, X., Davidson, E. A., Mauzerall, D. L., Searchinger, T. D., Dumas, P., & Shen, Y. (2015). Managing nitrogen for sustainable development. *Nature*, *528*, 51–59.

Zinkernagel, R., Evans, J., Neij, L. (2018). Applying the SDGs to cities: business as usual or a new dawn? *Sustainability*. *10*, 3201. https://doi.org/10.3390/su10093201

INDEX

Note: Page numbers followed by "*n*" indicate notes.

Printed and bound by CPI Group (UK) Ltd, Croydon, CR0 4YY

13/09/2026

14961060-0001